Holt McDougal Mathematics

Course 3
Chapter 7 Resource Book

HOLT McDOUGAL
a division of Houghton Mifflin Harcourt

Printed in the United States of America

ISBN 13: 978-0-55-400741-0
ISBN 10: 0-55-400741-X

3 4 5 170 12 11 10

Contents

Description of Contents

Family Involvement Pages

The Chapter Resource Book includes a set of family involvement pages for each section. The family involvement pages consist of the following items:

- **Family Letter**, a two-page letter describing the math that the student will study in each section. A list of vocabulary words from the lessons is included, and some worked-out examples are provided.
- **At-Home Practice**, a one-page worksheet with problems drawn from the content described in the Family Letter. Answers are provided on the page so that the student and his or her family can check this work at home.
- **Family Fun**, a one-page activity sheet that the student and his or her family can work on together.

Practice A, B, and C

There are three practice worksheets for every lesson. All of these reinforce the content of the lesson. Practice B is shown in the Teacher's Edition and is appropriate for the on-level student. It is also available as a workbook (Homework & Practice Workbook).

Practice A is easier than Practice B but still practices the content of the lesson. Practice C is more challenging than Practice B.

Review for Mastery

The Review for Mastery worksheet (one per lesson) provides an alternate way to teach or review the main concepts of the lesson. This worksheet is one or two pages long and is shown in the Teacher's Edition.

Challenge

The Challenge worksheet (one per lesson) enhances critical thinking skills and extends the lesson. This worksheet is shown in the Teacher's Edition.

Problem Solving

The Problem Solving worksheet (one per lesson) provides practice in problem solving and opportunities for real-world applications and for interdisciplinary connections. There are both multiple choice and short response problems. This worksheet is shown in the Teacher's Edition.

Reading Strategies

The Reading Strategies worksheet (one per lesson) provides tools to help the student master math vocabulary or symbols.

Puzzles, Twisters & Teasers

The Puzzles, Twisters & Teasers worksheet (one per lesson) provides fun practice while reinforcing the content of the lesson.

CHAPTER 7

Family Letter

7A Two-Dimensional Geometry

Dear Family,

In this chapter, the student will begin to study geometry. During this process, the student will learn to classify and name many geometrical figures. First, the student will review the foundations of geometry, which includes angle and line relationships.

Angles are very important when studying geometry. Special angles have special names and relationships. Knowing these relationships will allow the student to find unknown angle measurements.

Acute: Angles that measure less than 90°.

Obtuse: Angles that measure more than 90° but less than 180°.

Right: Angles that measure exactly 90°.

Complementary: Angles whose measures add to 90°.

Supplementary: Angles whose measures add to 180°.

The student will also learn about the angles formed by parallel lines and a transversal, and the relationships between them. In the figure below, line *m* || line *n*.

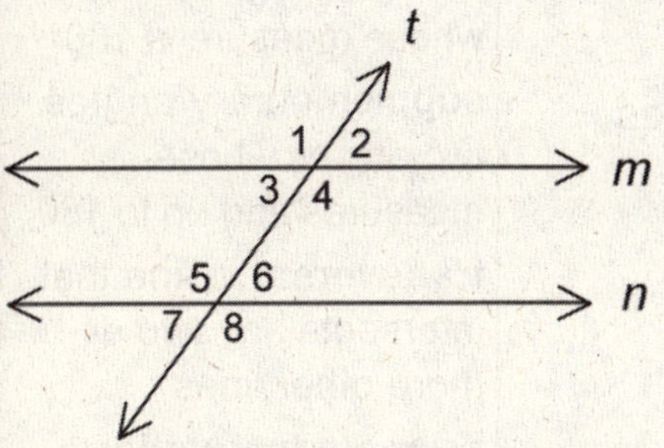

- Adjacent angles (e.g. $\angle 5$ and $\angle 6$) are supplementary.
- Vertical angles (e.g. $\angle 2$ and $\angle 3$) are congruent.
- Corresponding angles (e.g. $\angle 1$ and $\angle 5$) are congruent.
- Alternate interior angles (e.g. $\angle 3$ and $\angle 6$) are congruent.
- Alternate exterior angles (e.g. $\angle 2$ and $\angle 3$) are supplementary.

Vocabulary

These are the math words we are learning:

acute angle an angle whose measure is greater than 0° and less than 90°

acute triangle a triangle with 3 acute angles

adjacent angles angles that have a common vertex and side, but no common interior points

angle formed by two rays with a common endpoint called the vertex

complementary angles angles whose measures add up to 90°

congruent angles angles that have the same measure

equilateral triangle a triangle with 3 congruent sides

isosceles triangle a triangle with at least 2 congruent sides

midpoint the point that divides a segment into two congruent segments

obtuse angle an angle whose measure is greater than 90° and less than 180°

obtuse triangle a triangle with an obtuse angle

parallel lines lines in a plane that never meet

parallelogram a quadrilateral with 2 pairs of parallel sides

CHAPTER 7

Family Letter

7A Two-Dimensional Geometry continued

Polygons are also an important part of geometry. One special polygon is a triangle. The student will learn how to classify triangles by the angles or side lengths.

Triangle	Description	Picture
Acute Triangle	A triangle with 3 acute angles.	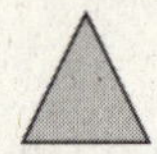
Right Triangle	A triangle with 1 right angle.	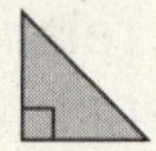
Obtuse Triangle	A triangle with 1 obtuse angle	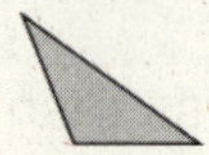
Equilateral Triangle	A triangle with 3 congruent sides and 3 congruent angles.	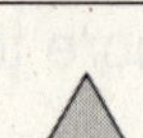
Isosceles Triangle	A triangle with 2 congruent sides and 2 congruent angles.	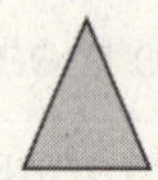
Scalene Triangle	A triangle with no congruent sides or congruent angles.	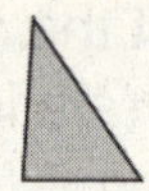

The student will use the ***Triangle Sum Theorem*** to find unknown angle measures in a triangle. The theorem states that the sum of the angles of any triangle equals 180°.

Find the angle measures in the isosceles triangle.

$68° + k° + k° = 180°$	Triangle Sum Theorem
$68° + 2k° = 180°$	Combine like terms.
$\underline{-68° \quad -68°}$	Subtract 68° from both sides.
$2k° = 112°$	
$\frac{2k°}{2} = \frac{112°}{2}$	Divide both sides by 2.
$k° = 56°$	

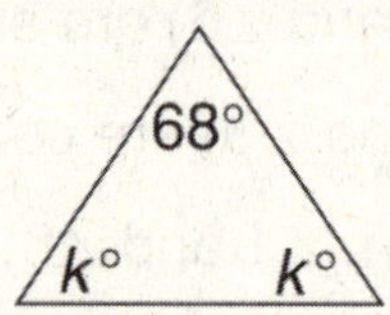

The triangle has angle measures of 56°, 56° and 68°.

The student will learn to graph polygons on a coordinate plane by placing vertices at different coordinates. He or she will learn to identify the missing vertex, given the other vertices of a polygon.

Sincerely,

perpendicular lines lines that intersect at 90° angles

polygon a closed plane figure formed by three or more segments

rectangle a quadrilateral with 4 right angles

regular polygon a polygon in which the sides and the angles have equal measures

right angle an angle whose measure is 90°

right triangle a triangle with a right angle

scalene triangle a triangle with no congruent sides

square a quadrilateral with 4 congruent sides and 4 right angles

straight angle an angle whose measure is 180°

supplementary angles two angles whose measures add up to 180°

transversal a line that intersects any two or more other lines

Triangle Inequality Theorem the sum of the lengths of any two sides of a triangle is greater than the length of the third side

Triangle Sum Theorem the angle measures of a triangle add up to 180°

vertical angles the nonadjacent angles formed by two intersecting lines

Name ______________________ Date ______________ Class ______________

CHAPTER 7

At-Home Practice

7A Two-Dimensional Geometry

1. m∠1 ______________________

2. m∠2 ______________________

3. m∠3 ______________________

4. m∠4 ______________________

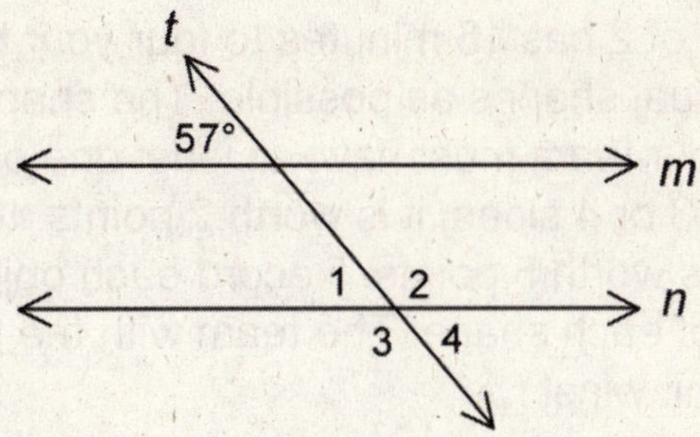

Find the value of each variable.

5.

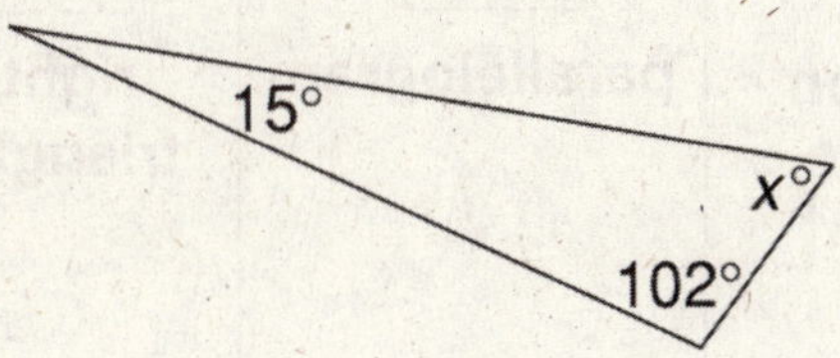

____

6.

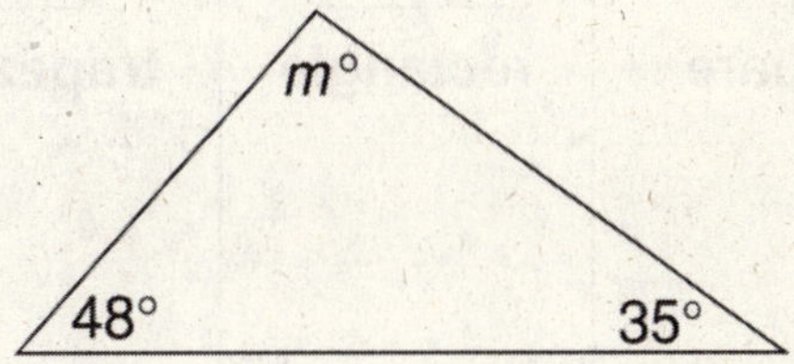

____

Write all the names that apply to each figure.

7.

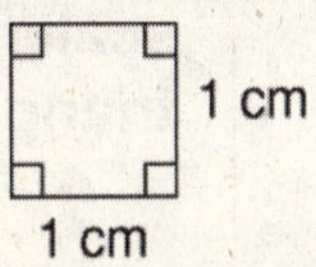

____

8.

____

9.

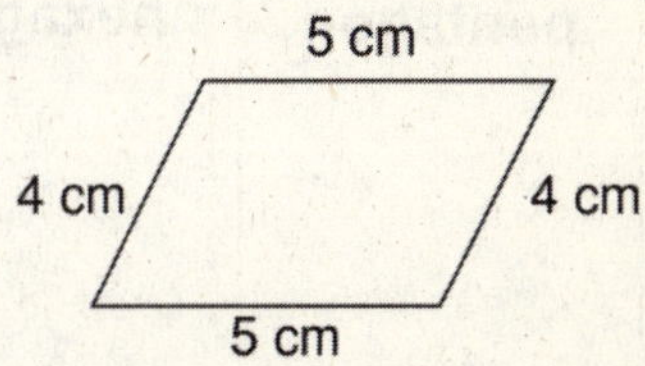

____

Find the sum of the angle measures in each figure.

10. regular octagon

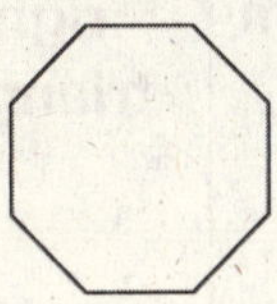

____

11. irregular hexagon

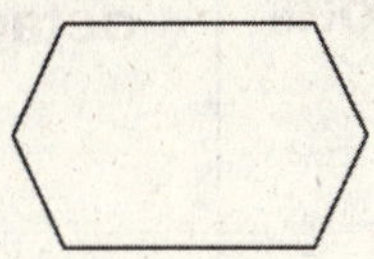

____

12. irregular pentagon

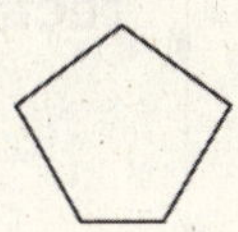

____

Answers: 1. 57° **2.** 123° **3.** 123° **4.** 57° **5.** 63° **6.** 97° **7.** quadrilateral, parallelogram, rectangle, rhombus, square **8.** quadrilateral, trapezoid **9.** quadrilateral, parallelogram **10.** 1080° **11.** 720° **12.** 540°

Name ______________________ Date ______________ Class ______________

CHAPTER 7

Family Fun

Scavenger Hunt

Directions

Each team of 2 has 15 minutes to tour your home and search for as many unusual shapes as possible. The shapes may be inside or outside. Your team must have at least one object per shape. If the shape has 3 or 4 sides, it is worth 2 points. If a shape has more than 4 sides, it is worth 5 points. Record each object your team finds under the name of each shape. The team with the most points at the end of the time limit wins!

square	rectangle	trapezold	octagon	parallelogram	right triangle
pentagon	hexagon	isosceles triangle	rhombus	equilateral triangle	scalene triangle

square	rectangle	trapezold	octagon	parallelogram	right triangle
pentagon	hexagon	isosceles triangle	rhombus	equilateral triangle	scalene triangle

Name ______________________ Date ______________ Class ______________

LESSON 7-1

Practice A

Angle Relationships

Use the diagram to name each figure.

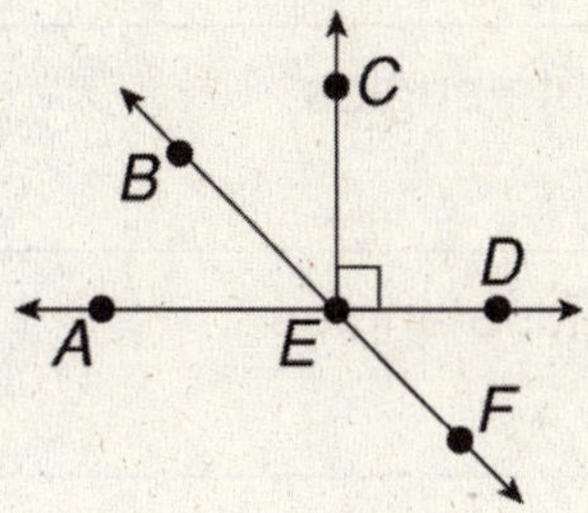

1. a right angle

____

2. two acute angles

____

3. two obtuse angles

____

____

4. pair of complementary angles

____

____

5. two pairs of supplementary angles

____

____

Use the diagram to find each angle measure.

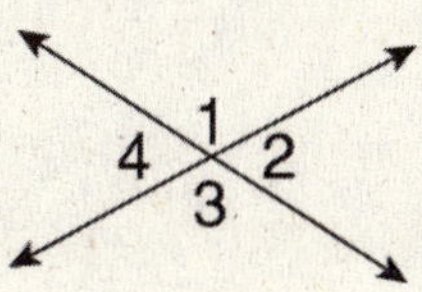

6. If m∠1 = 120°, find m∠3.

____

7. If m∠2 = x°, find m∠4.

____

8. If m∠3 = 110°, find m∠2.

____

9. If m∠4 = 65°, find m∠1.

____

Name ________________ Date ________ Class ________

LESSON 7-1

Practice B

Angle Relationships

Use the diagram to name each figure.

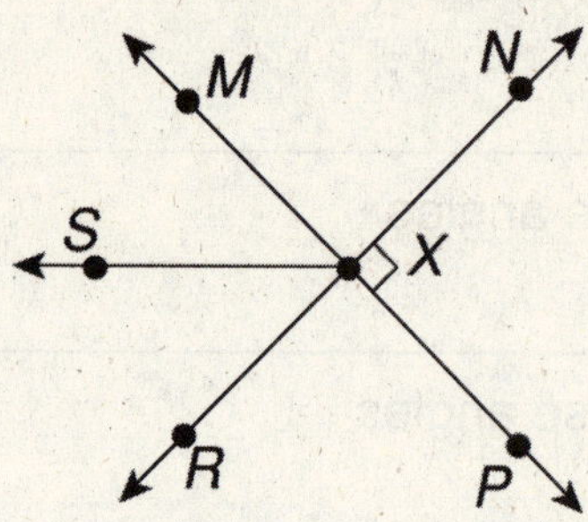

1. a right angle

____

2. two acute angles

____

3. two obtuse angles

____

4. a pair of complementary angles

____

____

____

5. three pairs of supplementary angles

____

____

____

Use the diagram to find each angle measure.

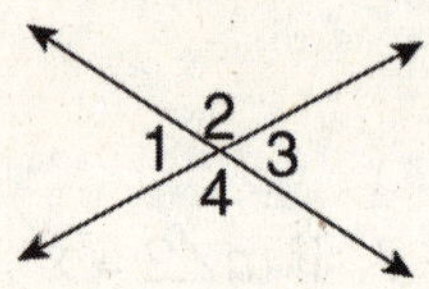

6. If m∠2 = 110°, find m∠4.

____

7. If m∠1 = *n*°, find m∠3.

____

8. The diagram shows the intersection of three roadways on a map. Based on the diagram, what should be the measure of ∠*ECD*?

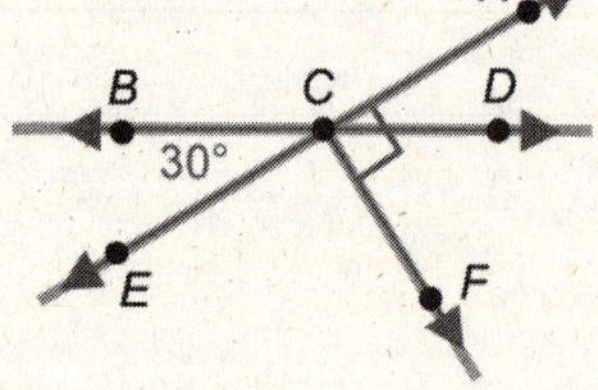

Name ______________________ Date ______________ Class ______________

LESSON 7-1

Practice C

Angle Relationships

Use the diagram to solve.

Write *true* or *false*. If a statement is false, rewrite it so it is true.

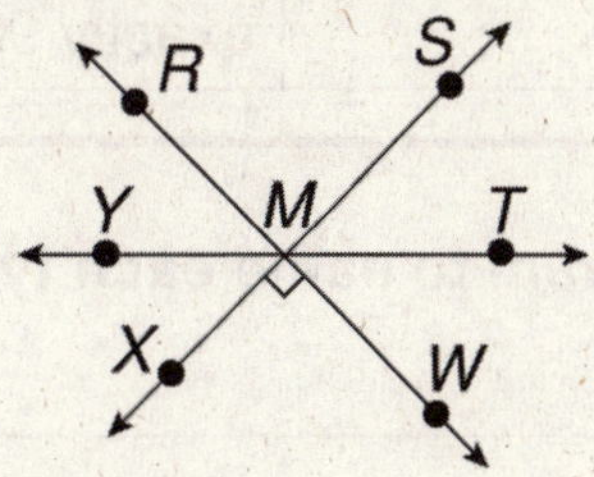

1. *RW* is a line.

____

2. Rays *YM* and *MS* make up line *YT*.

____

____

3. Angle *RMT* is an obtuse angle.

____

4. ∠*WMY* and ∠*RMY* are supplementary angles.

____

5. ∠*YMX* and ∠*SMT* are supplementary angles.

____

____

6. ∠*XMY* and ∠*YMR* are complementary angles.

____

____

7. If m∠*SMT* = 48°, then m∠*TMW* = 48°.

____

____

8. If m∠*SMY* = 135°, then m∠*RMT* = 135°.

____

____

9. If m∠*TMW* = *x*°, then m∠*RMT* = 180° − *x*°.

____

10. m∠*RMY* + m∠*SMT* + m∠*RMS* = 180°.

____

Use the diagram to find each angle measure.

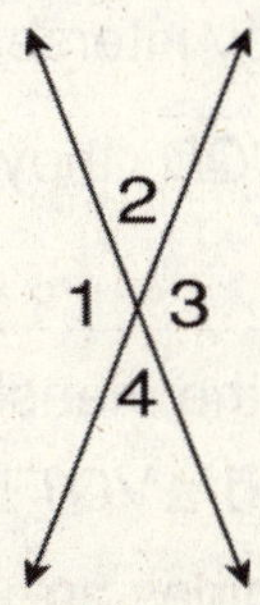

11. If m∠1 = 123°, find m∠3.

____

12. If m∠1 + m∠2 = 180° and m∠2 = 40°, find m∠3.

____

Name ______________________ Date ______________ Class ______________

LESSON 7-1

Review for Mastery

Angle Relationships

Acute Angle	Right Angle	Obtuse Angle	Straight Angle
Measures between 0º and 90º	Measures exactly 90º	Measures between 90º and 180º	Measures exactly 180º.

Use the diagram to name each type of angle.

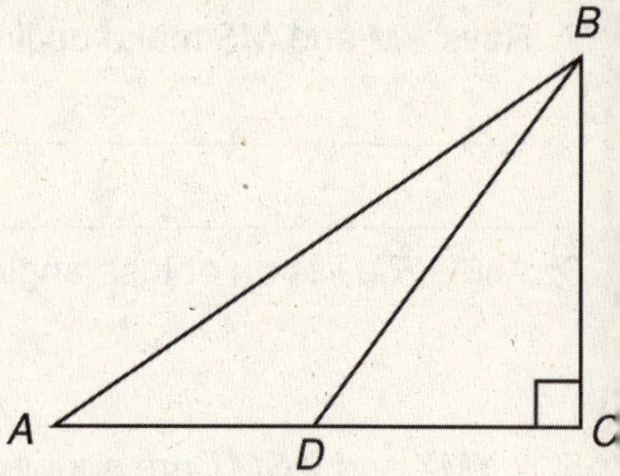

1. ∠*BCD* ______________________

2. ∠*BAD* ______________________

3. ∠*BDA* ______________________

4. ∠*CDA* ______________________

Complementary Angles	Supplementary Angles	Vertical Angles
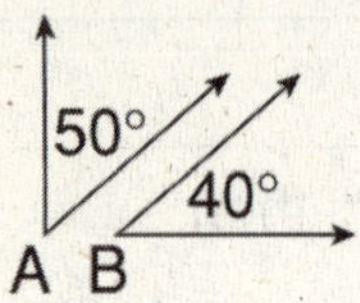	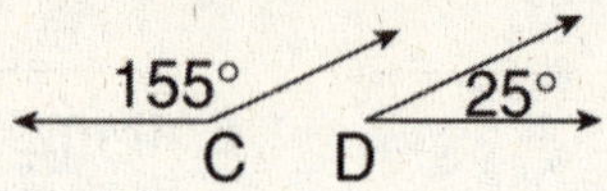	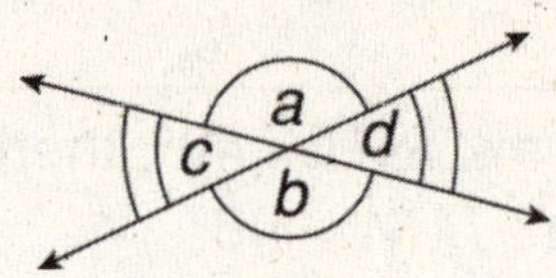
Two angles whose measures have a sum of 90º.	Two angles whose measures have a sum of 180º.	Intersecting lines form two pairs of vertical angles.

Use the diagram to complete the following.

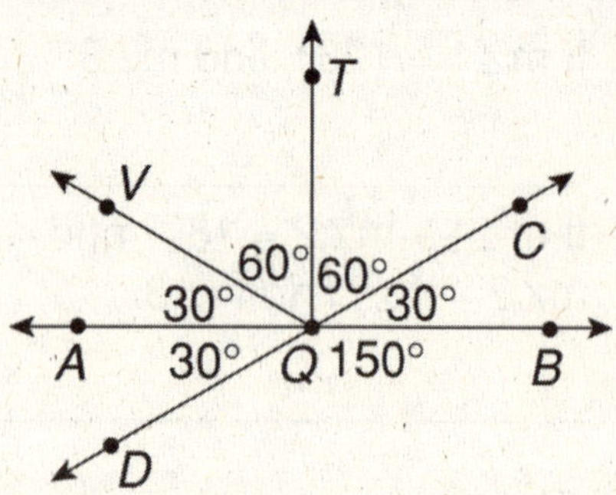

5. Since ∠*AQC* and ∠*DQB* are formed by intersecting lines, $\overleftrightarrow{AQB}$ and $\overleftrightarrow{CQD}$, they are:

   ____

6. The sum of the measures of ∠*AQV* and ∠*VQT* is: ________

   So, these angles are:

   ____

7. The sum of the measures of ∠*AQC* and ∠*CQB* is: ________

   So, these angles are: ______________________

Name ________________ Date ________________ Class ________________

LESSON 7-1

Challenge

Let's Meet!

Materials needed: paper strips, index cards, and scissors

1. Use a flat surface such as the top of your desk to represent a plane. Use strips of paper to represent lines. Move the lines around in the plane (**coplanar lines**) to determine the number of vertical angle pairs that are possible. Have all lines intersect at the same point. Summarize your results in a table.

Name of Coplanar Lines	Possible Number of Vertical Pairs
2	2
3	
4	
5	

2. Slit one index card and connect two cards to model two intersecting planes.

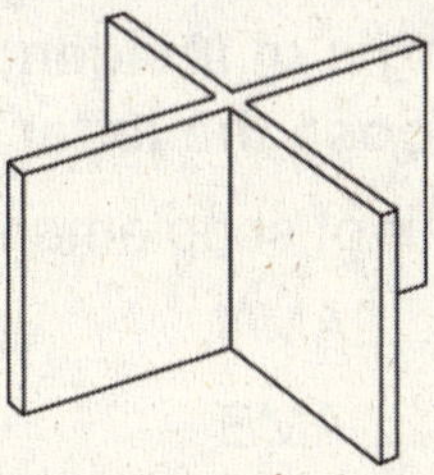

a. How many pairs of supplementary angles are there?

____

b. Name the supplementary angles.

____

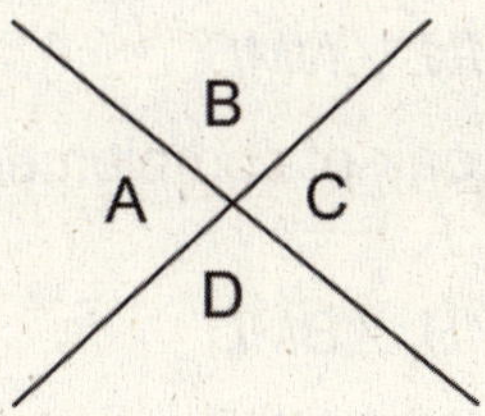

c. If m∠A = 90°, what is the measure of the other angles?

____

Name ______________________ Date ______________ Class ______________

LESSON 7-1

Problem Solving

Angle Relationships

Use the flag of the Bahamas to solve the problems.

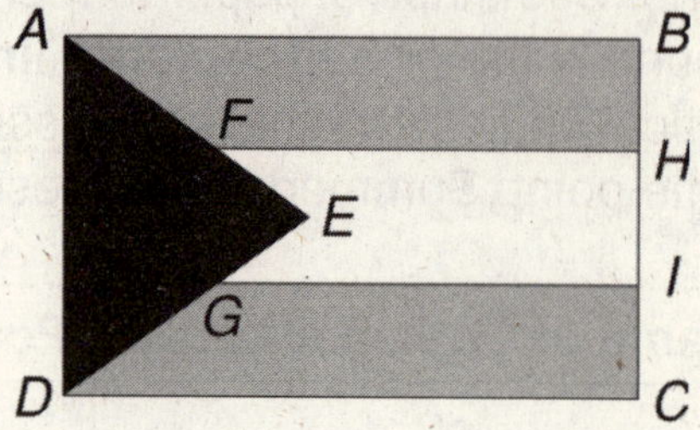

1. Name a right angle in the flag.

____

2. Name a pair of complementary angles in the flag.

____

3. Name two acute angles in the flag.

____

4. Name a pair of supplementary angles in the flag.

____

The diagram illustrates a ray of light being reflected off a mirror. The angle of incidence is congruent to the angle of reflection. Choose the letter for the best answer.

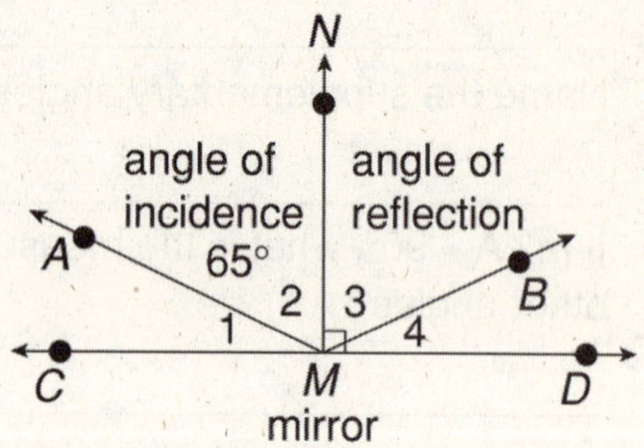

5. Name a pair of supplementary angles.
   - A ∠CMA, ∠AMN,
   - B ∠NMB, ∠AMB
   - C ∠NMC, ∠DMN
   - D ∠AMN, ∠NMD

6. Name a pair of complementary angles.
   - F ∠NMB, ∠BMD
   - G ∠AMN, ∠NMB
   - H ∠CMA, ∠AMD
   - J ∠CMA, ∠DMB

7. Which angle is congruent to ∠2?
   - A ∠1
   - B ∠4
   - C ∠3
   - D none

8. Find the measure of ∠4.
   - F 65°
   - G 35°
   - H 25°
   - J 90°

9. Find the measure of ∠1.
   - A 65°
   - B 35°
   - C 25°
   - D 90°

10. Find the measure of ∠3.
   - F 90°
   - G 45°
   - H 35°
   - J 65°

Name ______________________ Date ______________ Class ______________

LESSON 7-1

Reading Strategies

Understanding Vocabulary

An **angle** is a figure formed by two rays, lines or line segments with a common point. There are multiple ways to name an angle.

- Name a point on one line, the vertex or point of intersection, then a point on the other line.
  $\angle ABC$ or $\angle CBA$
- The vertex can name the angle.
  $\angle B$ or $\angle 1$

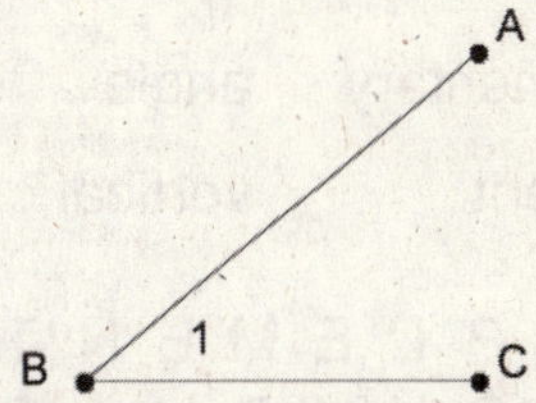

There are four types of angles, an **acute angle**, an **obtuse angle**, a **right angle**, and a **straight angle**.

Acute angle	Obtuse angle	Right angle	Straight angle.
The measure is greater than 0º and less than 90º.	The measure is greater than 90º and less than 180º.	The measure is 90º.	The measure is 180º.

Use this figure to answer each question.

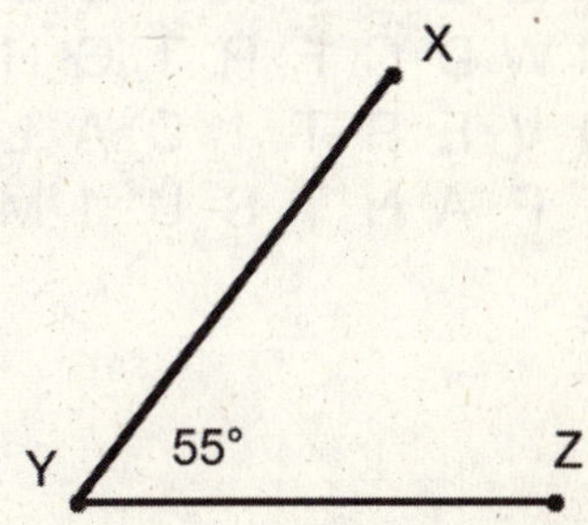

1. How would you name the angle?

   ____

2. Classify the angle.

   ____

Pairs of angles can also be classified into different types.

Complementary angles	Supplementary angles	Adjacent angles	Vertical angles	Congruent angles
The angle measures equal 90º.	The angle measures equal 180º.	The angles share a common vertex and side.	The angles are nonadjacent angles formed by two intersecting lines.	The angles have the same measure.

3. In the figure above, what angle measure would be complementary to the one shown?

   ____

Name ______________________ Date ____________ Class ____________

LESSON 7-1

Puzzles, Twisters & Teasers

What's Your Angle?

Circle the words below in the word search (horizontally, vertically or diagonally). Then find a word in the word search that answers the riddle. Circle it and write it on the line.

supplementary angle complementary

congruent vertical

```
C O M P L E M E N T A R Y S
O M H O Y U P U I O N W E U
N O L A T Y L T R Y G R T P
G I U N R T A W E R L U I P
R A Y T Q W N P L M E Y H L
U W D V G T S Q W E R T Y E
E E D C V F R T G B R H Y M
N O I U Y T R F V B C J K E
T I S E G M A N T U Y T F N
I M L P O K N B J L U H V T
M W D C F R T G H N I J U A
Q V E R T I C A L D C N H R
E P A N T S U J M K I O B Y
```

Why did the fireman wear red suspenders?

To keep his ____ ____ ____ ____ ____ on.

Name ______________________ Date ______________ Class ______________

LESSON 7-2

Practice A

Parallel and Perpendicular Lines

1. Measure the angles formed by the transversal and the parallel lines. Which angles seem to be congruent?

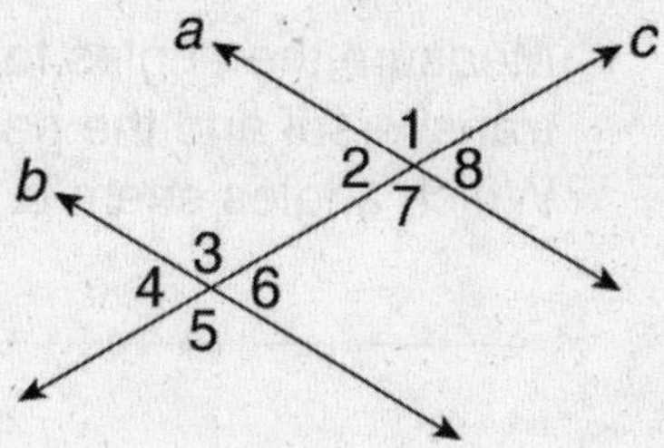

____

____

In the figure, line *r* || line *s*. Find the measure of each angle.

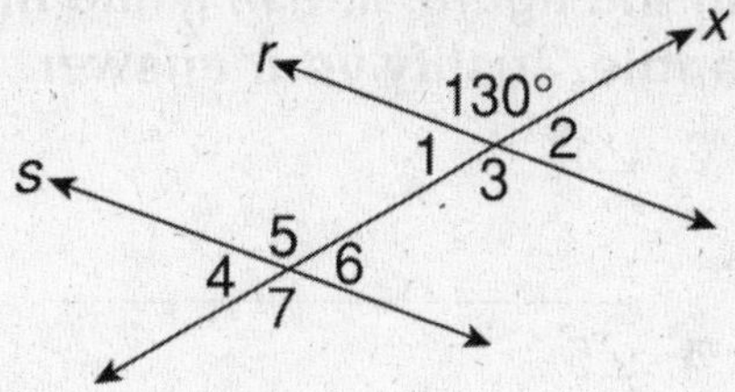

2. ∠1 ______ 3. ∠5 ______ 4. ∠6 ______

5. ∠7 ______ 6. ∠4 ______ 7. ∠3 ______

In the figure, line *m* || line *n*. Find the measure of each angle.

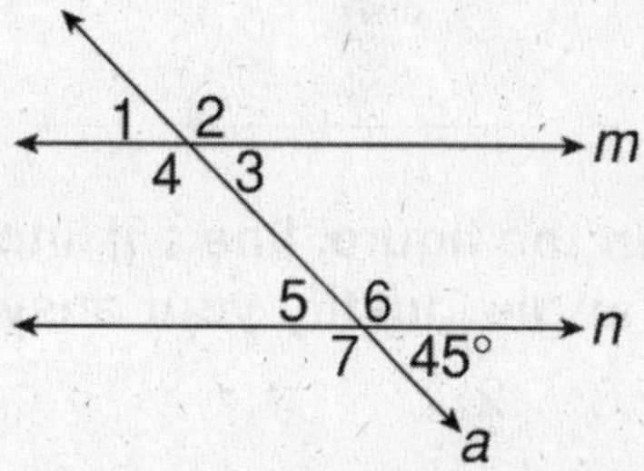

8. ∠1 ______ 9. ∠2 ______ 10. ∠3 ______

11. ∠5 ______ 12. ∠6 ______ 13. ∠7 ______

In the figure, line *a* || line *b*.

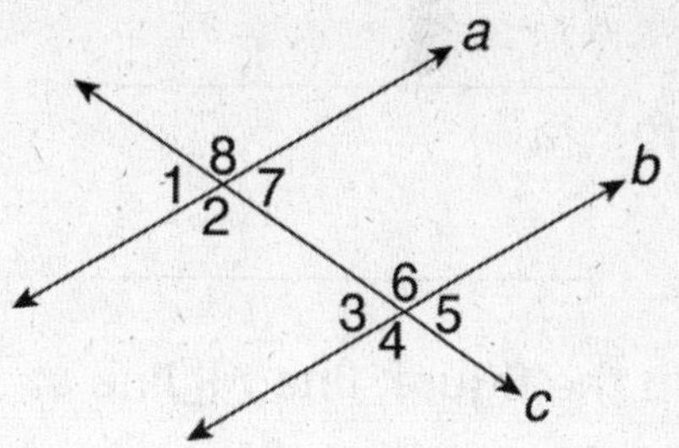

14. Name all angles congruent to ∠1.

____

15. Name all angles congruent to ∠2.

____

16. Name three pairs of supplementary angles.

____

17. Which line is the transversal?

____

Name ____________________ Date ______________ Class ______________

LESSON 7-2

Practice B

Parallel and Perpendicular Lines

1. Measure the angles formed by the transversal and the parallel lines. Which angles seem to be congruent?

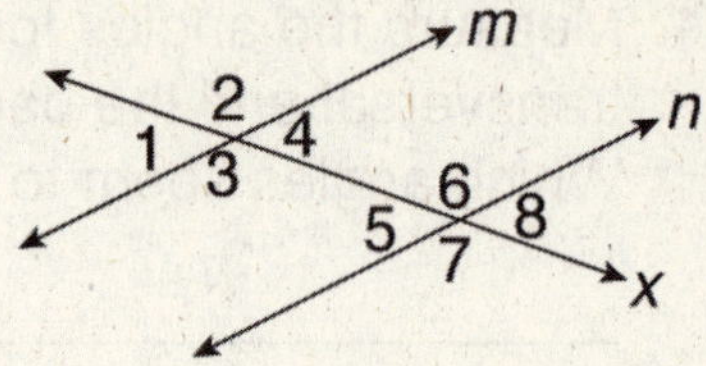

____

____

In the figure, line *m* || line *n*. Find the measure of each angle. Justify your answer.

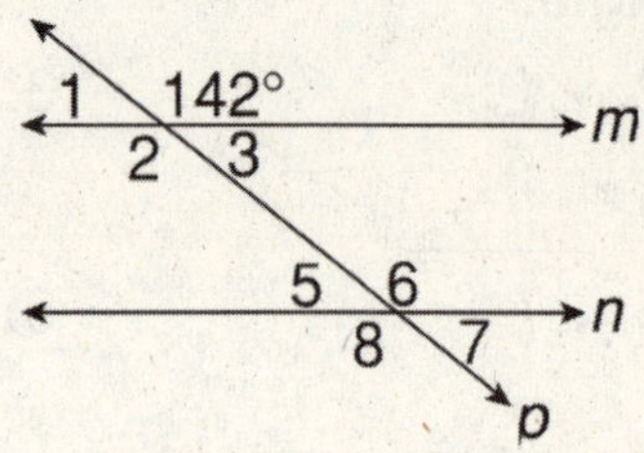

2. ∠1 ______________ 3. ∠2 ______________

4. ∠5 ______________ 5. ∠6 ______________

6. ∠8 ______________ 7. ∠7 ______________

In the figure, line *a* || line *b*. Find the measure of each angle. Justify your answer.

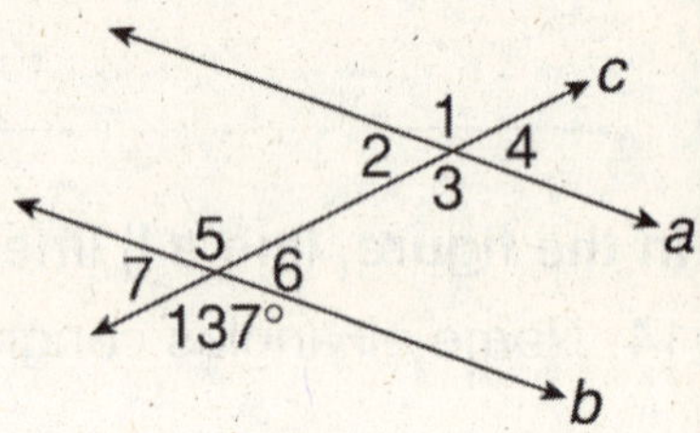

8. ∠2 ______________ 9. ∠5 ______________

10. ∠6 ______________ 11. ∠7 ______________

12. ∠4 ______________ 13. ∠3 ______________

In the figure, line *r* || line *s*.

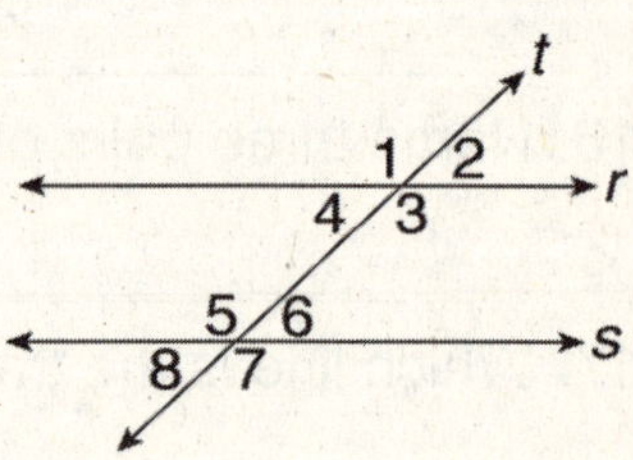

14. Name all angles congruent to ∠2.

____

15. Which line is the transversal?

____

Name ______________________ Date ______________ Class ______________

LESSON 7-2

Practice C

Parallel and Perpendicular Lines

In the figure, line *a* || line *b*.

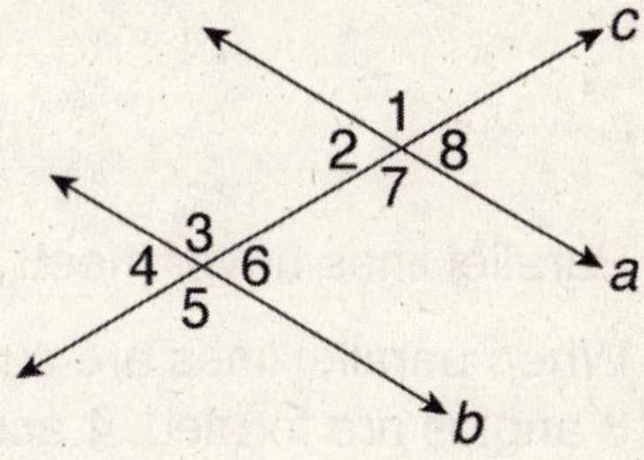

1. Name all angles congruent to ∠1.

____

2. Name all angles congruent to ∠2.

____

3. Name three pairs of supplementary angles.

____

4. Which line is the transversal?

____

5. If m∠7 is 131°, what is the m∠8?

____

6. If m∠4 is 57°, what is the m∠5?

____

7. If m∠3 is 127°, what is the m∠8?

____

8. If a transversal were drawn perpendicular to line *a* and line *b*, what would be the measure of the angles formed?

____

Draw a diagram to illustrate each of the following.

9. line *x* || line *y* || line *z* and transversal *k*

10. line r || line *s* and transversal *m* with eight congruent angles

11. line *a* || line *b* and transversal *t* with ∠1 ≅ ∠2 and ∠3 ≅ ∠4

Name ______________________ Date ______________ Class ______________

LESSON 7-2

Review for Mastery

Parallel and Perpendicular Lines

Parallel Lines

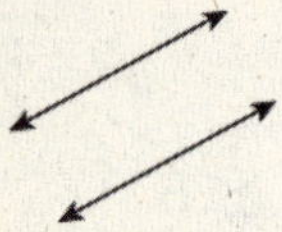

Parallel lines never meet.

Perpendicular Lines

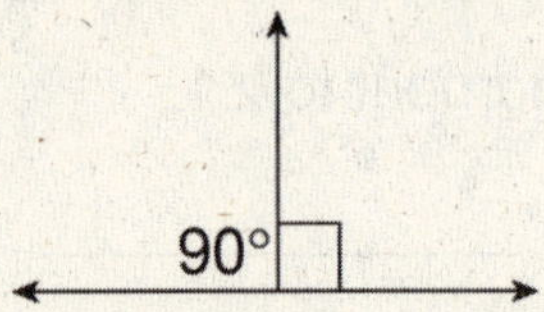

Perpendicular lines form right angles.

When parallel lines are cut by a **transversal**, 8 angles are formed, 4 acute and 4 obtuse.

The acute angles are all congruent.

The obtuse angles are all congruent.

Any acute angle is supplementary to any obtuse angle.

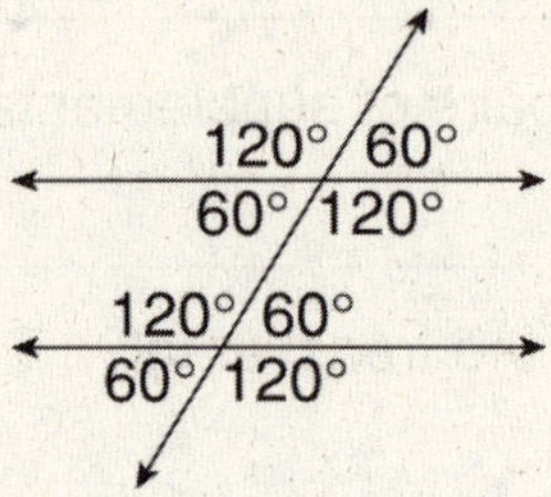

In each diagram, parallel lines are cut by a transversal. Name the angles that are congruent to the indicated angle.

1.

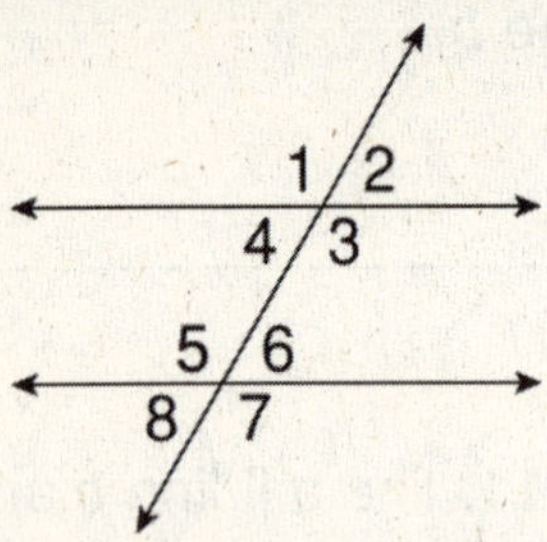

The angles congruent to ∠1 are:

____

2.

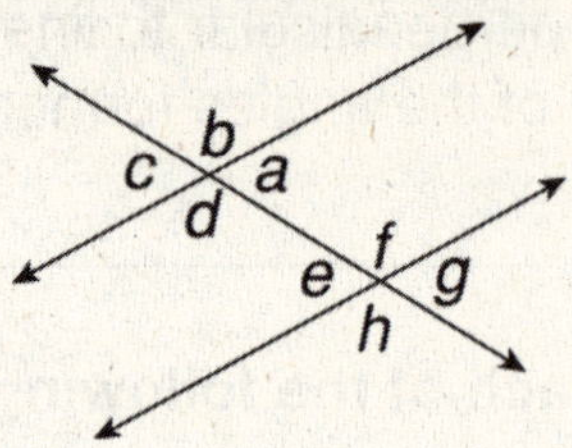

The angles congruent to ∠*a* are:

____

3.

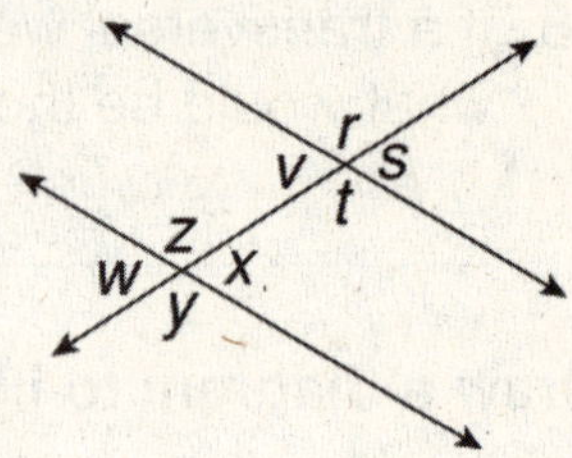

The angles congruent to ∠*z* are:

____

In each diagram, parallel lines are cut by a transversal and the measure of one angle is given. Write the measures of the remaining angles on the diagram.

4.

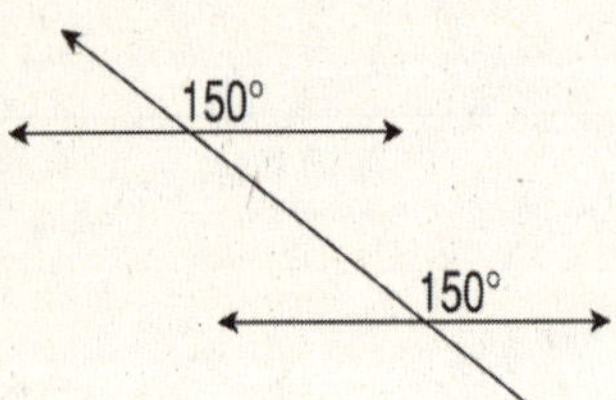

5.

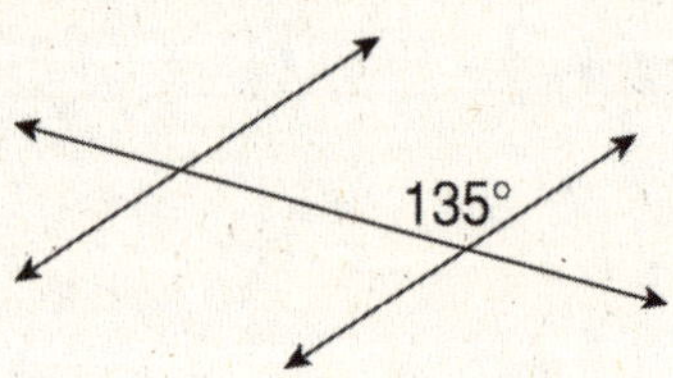

6.

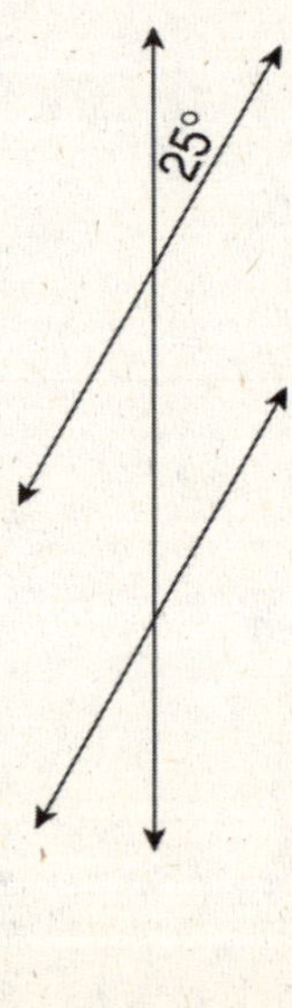

Name ______________________ Date ______________ Class ______________

LESSON 7-2

Challenge

Pairing Off

When two parallel lines are cut by a transversal, eight angles are formed. Of these, four angles are between the parallel lines, **interior angles.**

1. In this diagram, name the four interior angles formed by the parallel lines and the transversal.

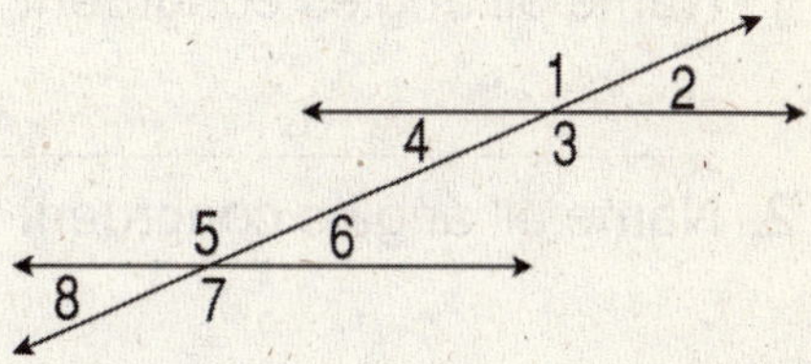

____

2. Think of the interior angles in pairs. Name the two pairs of interior angles that are on opposite sides of the transversal.

____

3. What is true about the measures of ∠3 and ∠5 in the diagram above? in the diagram at the right? Use a protractor to verify your conjecture.

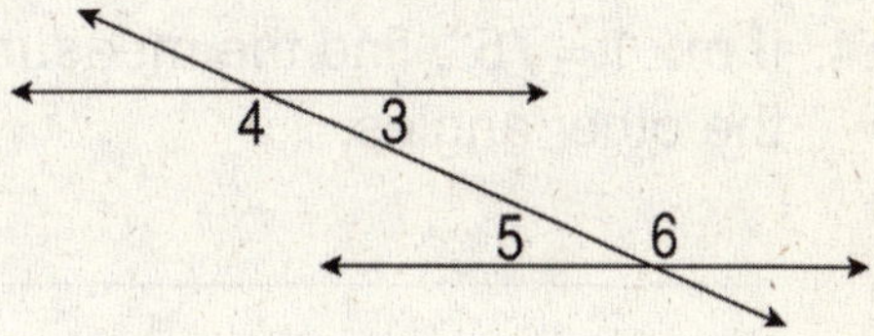

____

4. What is true about the measures of ∠4 and ∠6 in the diagram above? in the diagram at the right? Use a protractor to verify your conjecture.

____

5. Interior angles that are on opposite sides of the transversal are called **alternate interior angles.**

   Draw a conclusion about the measures of alternate interior angles formed by parallel lines and a transversal.

____

____

Use your observation about the measures of alternate interior angles of parallel lines to find the measure of ∠*x* in each of these diagrams.

6.

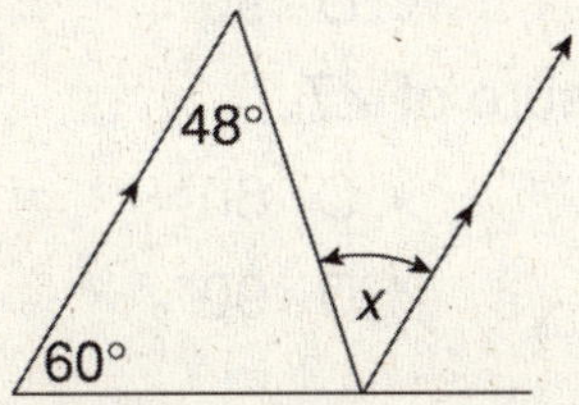

m∠*x* = ______________

7.

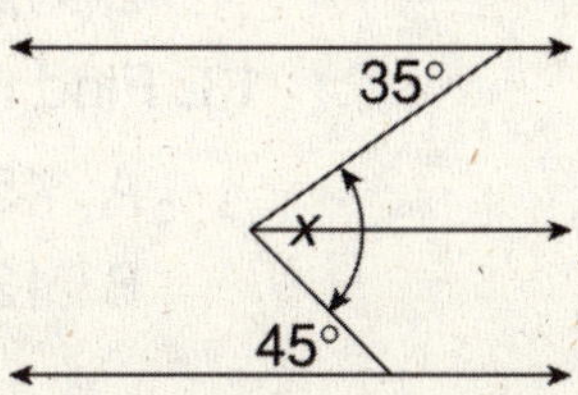

m∠*x* = ______________

8.

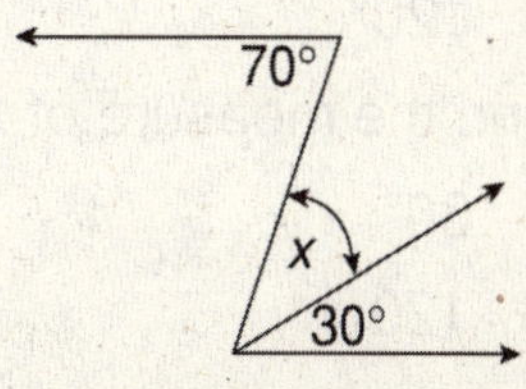

m∠*x* = ______________

Name ______________________ Date ______________ Class ______________

LESSON 7-2

Problem Solving

Parallel and Perpendicular Lines

The figure shows the layout of parking spaces in a parking lot.
$\overline{AB} \parallel \overline{CD} \parallel \overline{EF}$

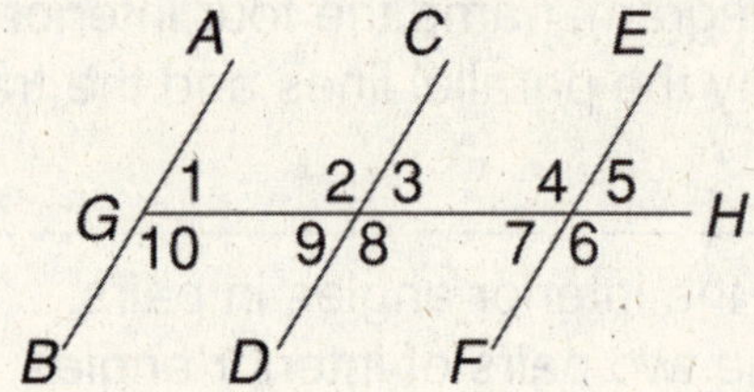

1. Name all angles congruent to ∠1.

____

2. Name all angles congruent to ∠2.

____

3. Name a pair of supplementary angles.

____

4. If m∠1 = 75°, find the measures of the other angles.

____

____

____

5. Name a pair of vertical angles.

____

6. If m∠1 = 90°, then $\overline{GH}$ is perpendicular to

____

The figure shows a board that will be cut along parallel segments *GB* and *CF*. $\overline{AD} \parallel \overline{HE}$. Choose the letter for the best answer.

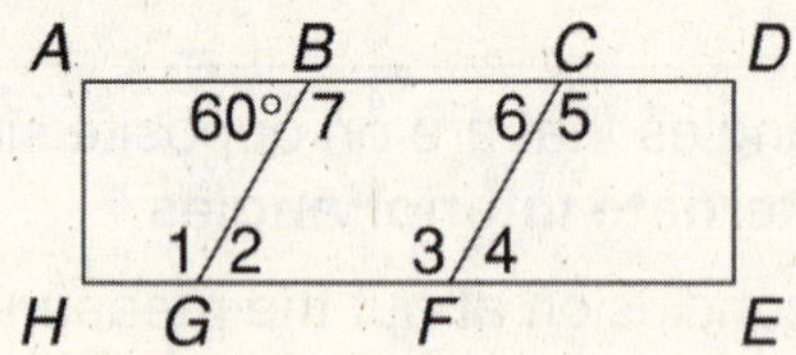

7. Find the measure of ∠1.
   A 45° C 60°
   B 120° D 90°

8. Find the measure of ∠2.
   F 30° H 60°
   G 120° J 90°

9. Find the measure of ∠3.
   A 30° C 60°
   B 120° D 90°

10. Find the measure of ∠4.
   F 45° H 60°
   G 120° J 90°

11. Find the measure of ∠5.
   A 30° C 60°
   B 120° D 90°

12. Find the measure of ∠6.
   F 30° H 60°
   G 120° J 90°

13. Find the measure of ∠7.
   A 45° C 60°
   B 120° D 90°

Name ______________________ Date ______________ Class ______________

LESSON 7-2

Reading Strategies

Understanding Symbols

The symbol || stands for **parallel**. Parallel lines are the same distance apart and never meet.

The yard lines marked on a football field are similar to parallel line segments.

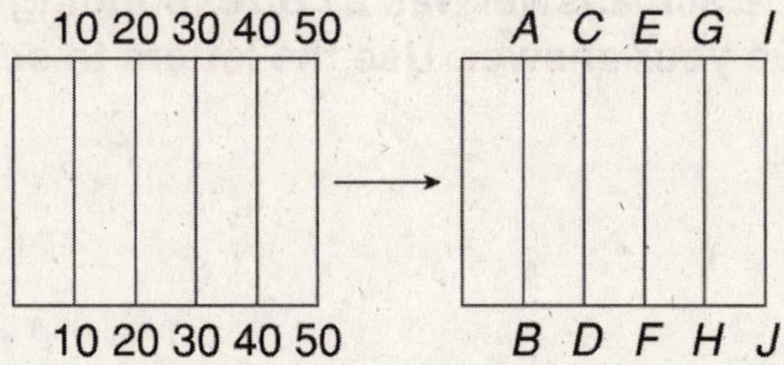

Segment *AB* is parallel to segment *CD*. This can be written: $\overline{AB} \parallel \overline{CD}$.

1. Identify another pair of parallel line segments in the figure above.

   ____

2. Use the || symbol to write how the line segments are related.

   ____

The symbol ⊥ stands for **perpendicular**. Perpendicular lines meet to form a square corner.

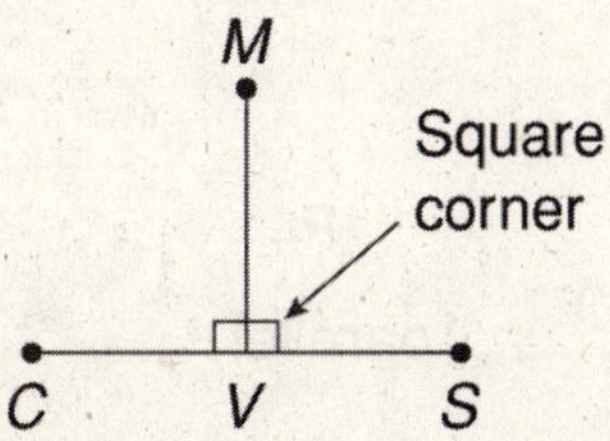

MV ⊥ CS

Read: segment *MV* is perpendicular to segment *CS*

A square in a corner is used as a symbol in a figure to show that lines are perpendicular.

Use this figure for Exercises 3–5.

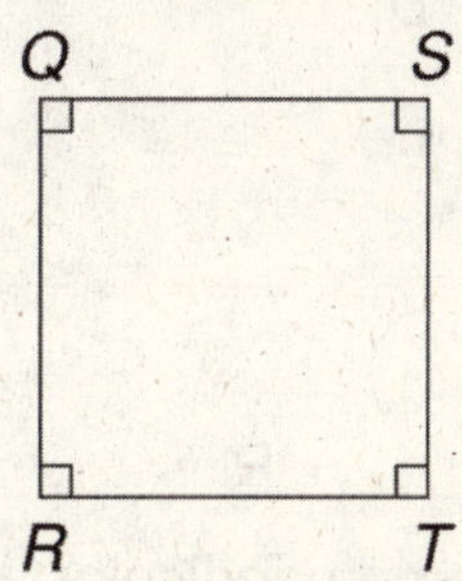

3. Segment *QR* is perpendicular to segment *TR*. Write this with symbols.

   ____

4. Identify two other line segments that are perpendicular to each other.

   ____

5. What symbol in the figure tells you that line segments are perpendicular to each other?

   ____

Name ________________________ Date ________________ Class ________________

LESSON 7-2

Puzzles, Twisters & Teasers

Line Up!

Decide whether the lettered lines in each figure are parallel or perpendicular. Each answer has a corresponding letter. Circle the letter above your answer. Use the letters to solve the riddle.

1.

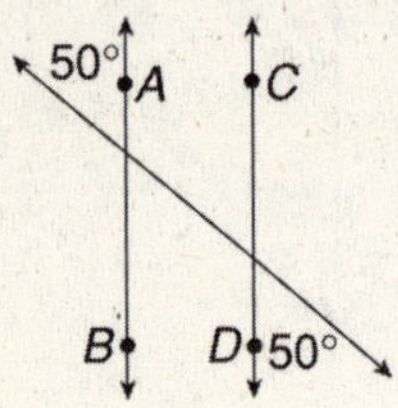

R — parallel   J — perpendicular

2.

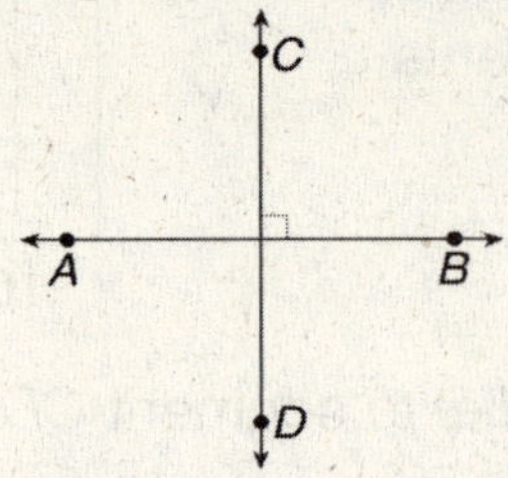

U — parallel   E — perpendicular

3.

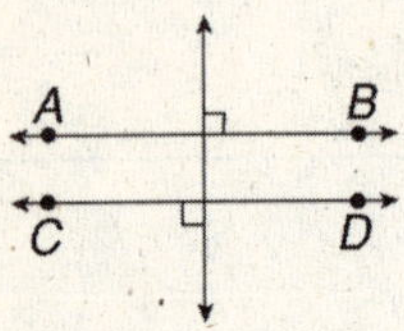

C — parallel   T — perpendicular

4.

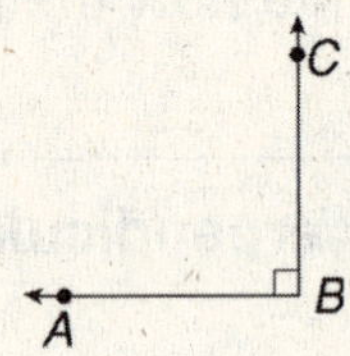

P — parallel   C — perpendicular

5.

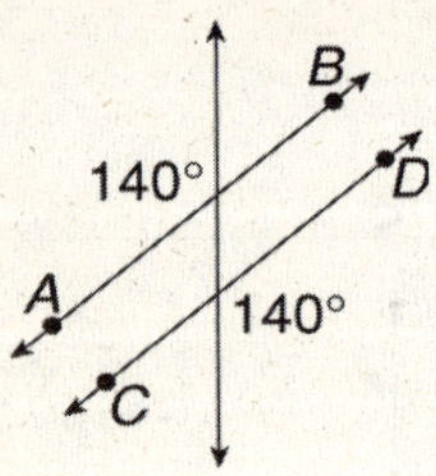

I — parallel   R — perpendicular

6.

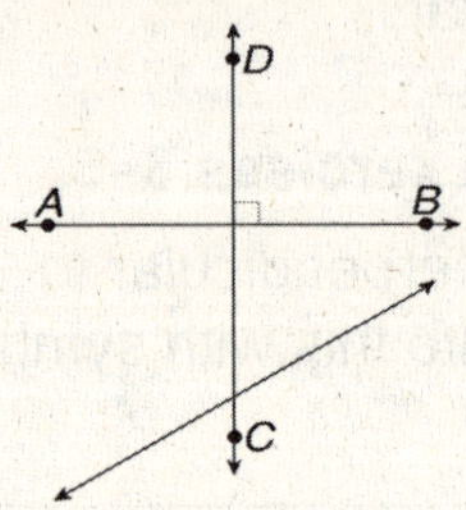

O — parallel   N — perpendicular

What do you call it when you ride your bike around and around the block?

___ ___ ___ Y ___ L ___ ___ G

1 2 3 4 5 6

Name ______________________ Date ______________ Class ______________

LESSON 7-3

Practice A

Triangles

Identify each triangle by its angles and sides.

1.

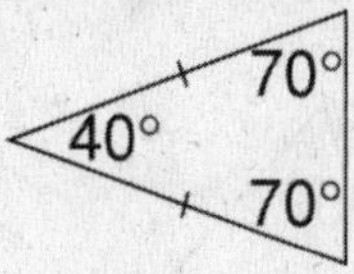

____

2.

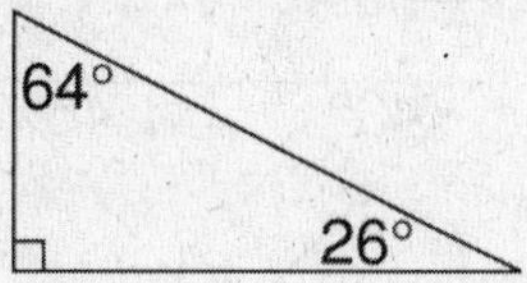

____

3.

____

Find each angle measure.

4. Find $x°$ in the acute triangle.

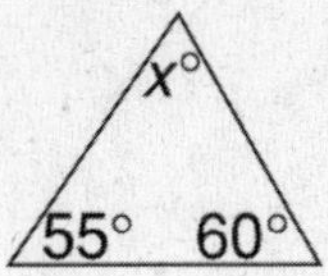

____

5. Find $y°$ in the right triangle.

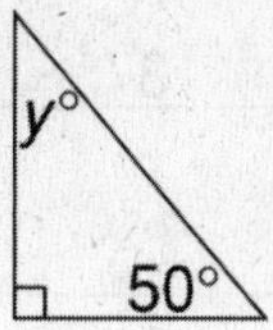

____

6. Find $r°$ in the obtuse triangle.

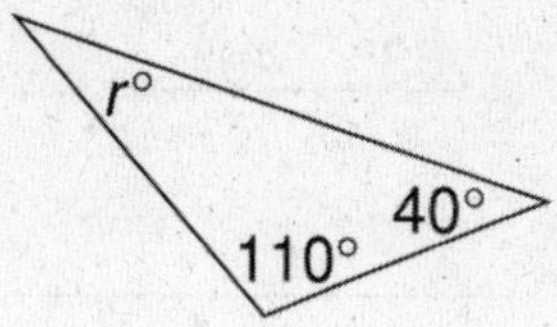

____

7. Find $x°$ in the acute triangle.

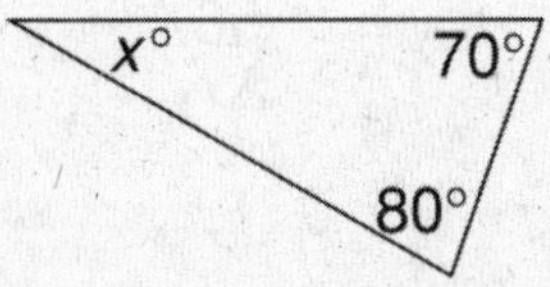

____

8. Find $y°$ in the right triangle.

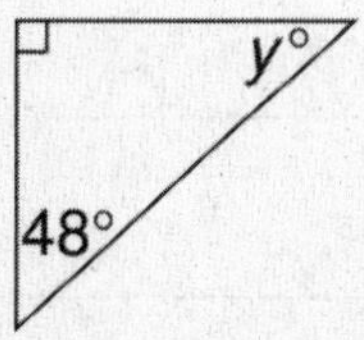

____

9. Find $m°$ in the obtuse triangle.

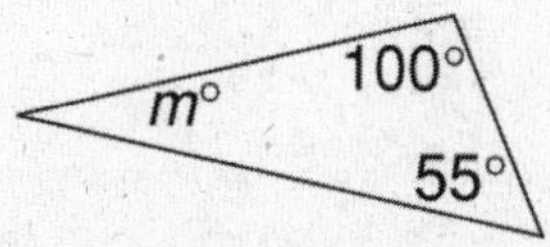

____

10. Find $t°$ in the isoceles triangle.

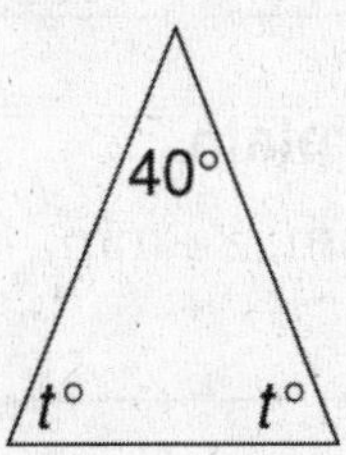

____

11. Find $x°$ in the scalene triangle.

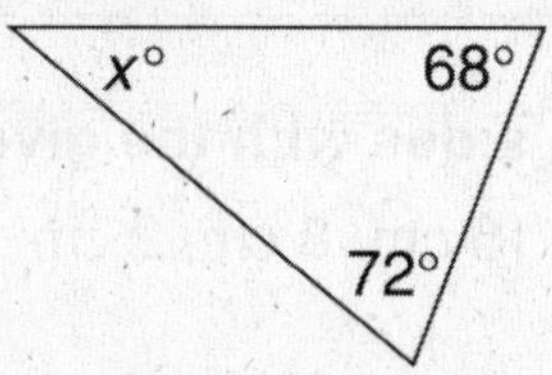

____

12. Find $n°$ in the isoceles triangle.

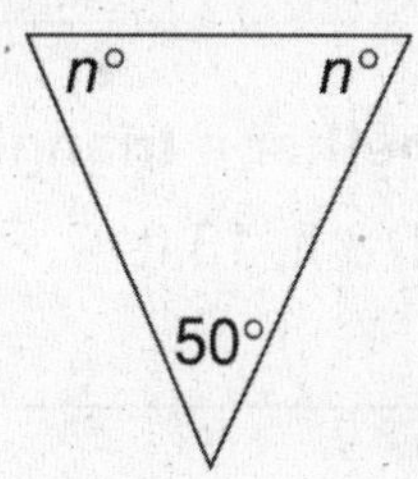

____

13. The second angle in a triangle is three times as large as the first. The third angle is one-third as large as the second angle. Find the angle measures. Draw a possible picture of the triangle.

____

Name ____________________ Date ____________ Class ____________

LESSON 7-3

Practice B

Triangles

1. Find $x°$ in the right triangle.

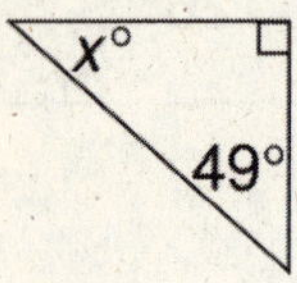

____

2. Find $y°$ in the obtuse triangle.

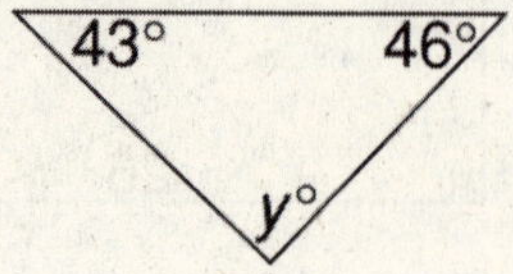

____

3. Find $m°$ in the acute triangle.

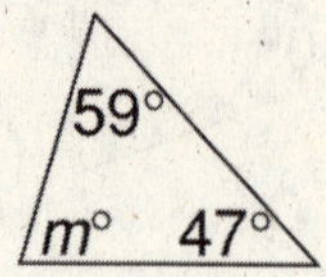

____

4. Find $w°$ in the acute triangle.

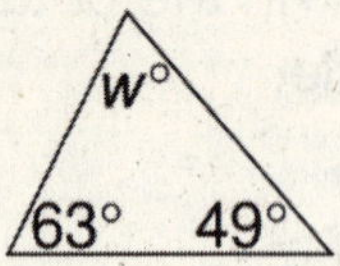

____

5. Find $t°$ in the scalene triangle.

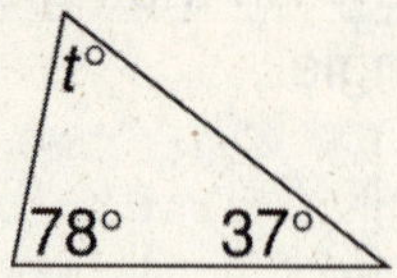

____

6. Find $n°$ in the scalene triangle.

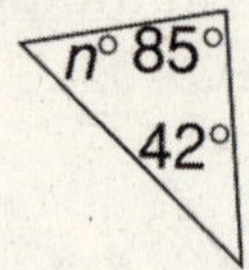

____

7. Find $x°$ in the isosceles triangle.

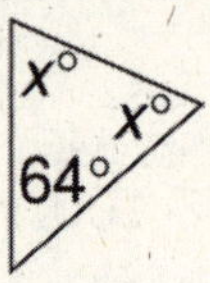

____

8. Find y in the equilateral triangle.

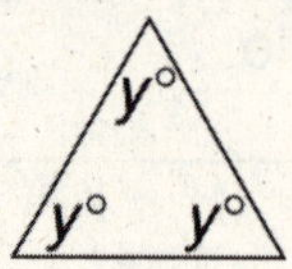

____

9. Find r in the isosceles triangle.

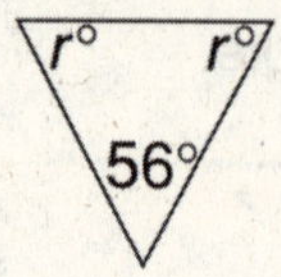

____

10. The second angle in a triangle is one third as large as the first. The third angle is two thirds as large as the first angle. Find the angle measures. Draw a possible picture of the triangle.

____

Tell whether a triangle can have sides with the given lengths. Explain.

11. 6 ft, 8 ft, 13 ft

____

12. 15 cm, 8 cm, 2 cm

____

13. 9 mm, 22 mm, 14 mm

____

Name ______________________ Date ______________ Class ______________

LESSON 7-3

Practice C

Triangles

Find the value of each variable.

1.

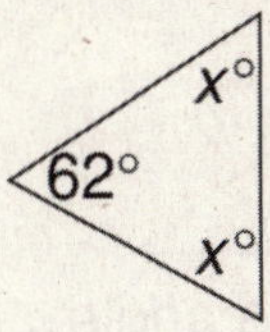

____

2.

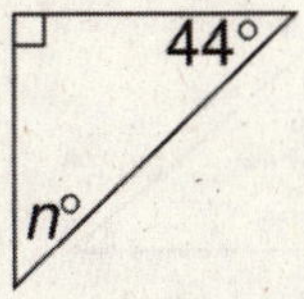

____

3.

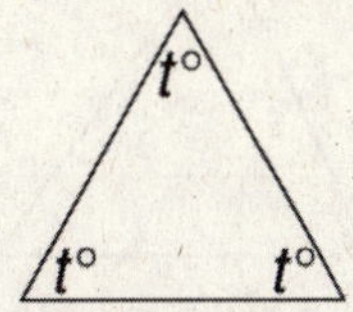

____

4.

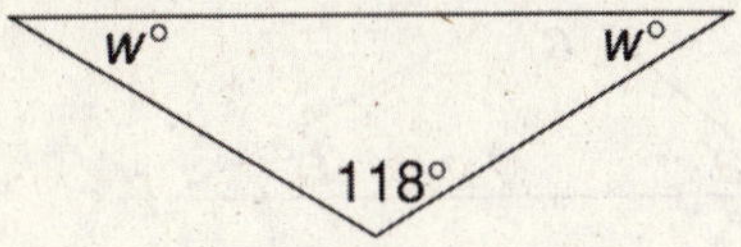

____

5.

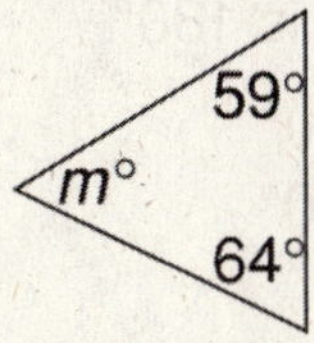

____

6.

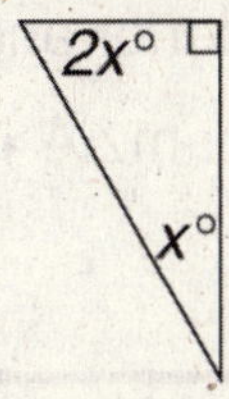

____

7. The measure of the second angle in a triangle is four more than the measure of the first angle and the measure of the third angle is eight more than twice the measure of the first angle. Find the measure of each angle.

____

Describe each statement as *always*, *sometimes*, or *never* true.

8. An obtuse triangle is a scalene triangle.

____

9. A scalene triangle is an isosceles triangle.

____

10. A right triangle is an isosceles triangle.

____

11. A triangle with all angles congruent is acute.

____

Sketch a triangle to fit each description. If no triangle can be drawn, write *not possible*.

12. obtuse, isosceles

13. right, scalene

14. acute, equilateral

Name ______________________ Date ____________ Class ____________

LESSON 7-3

Review for Mastery

Triangles

Acute Triangle

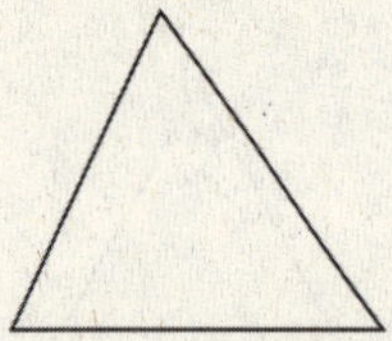

3 acute angles

Right Triangle

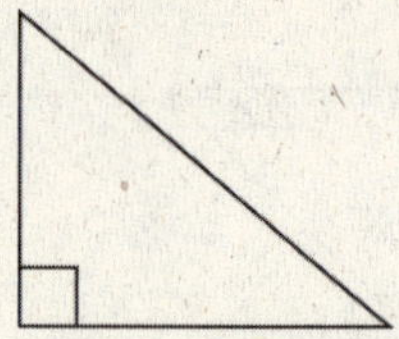

1 right angle

Obtuse Triangle

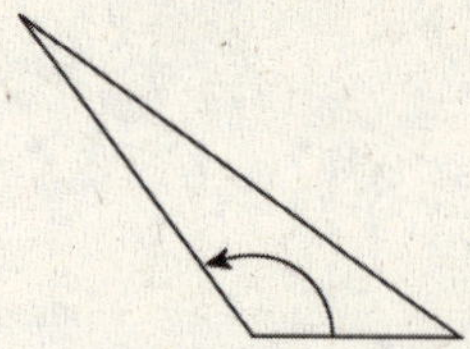

1 obtuse angle

Triangle Sum Theorem: The sum of the measures of the three interior angles of any triangle is 180°.

In the diagram: $m\angle A + m\angle B + m\angle C = 180°$

$a° + b° + c° = 180°$

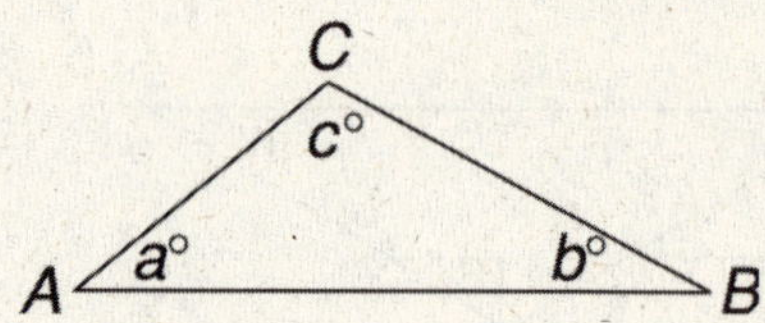

Complete to find the measure of the unknown angle.

1.

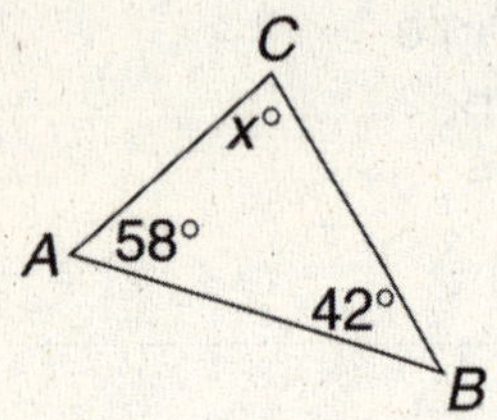

$m\angle A + m\angle B + m\angle C = 180°$

_______ + _______ + $x°$ = _______

______________ + $x°$ = _______

$x°$ = _______

2.

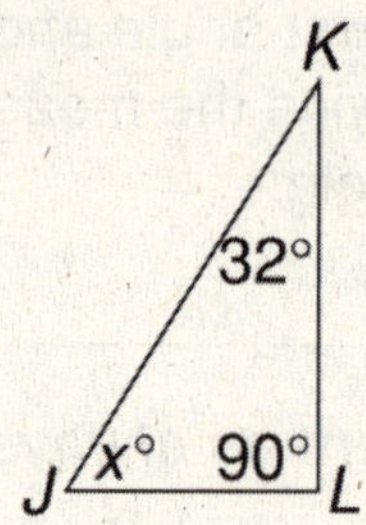

$m\angle J + m\angle K + m\angle L = 180°$

$x°$ + _______ + _______ = _______

$x°$ + ______________ = _______

$x°$ = _______

3.

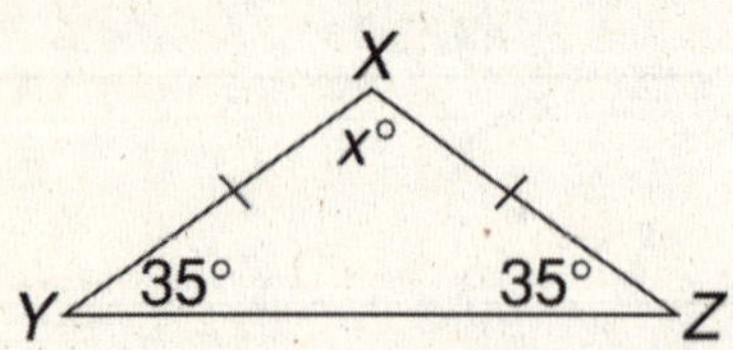

$m\angle X + m\angle Y + m\angle Z = 180°$

____

____

$x°$ = _______

4.

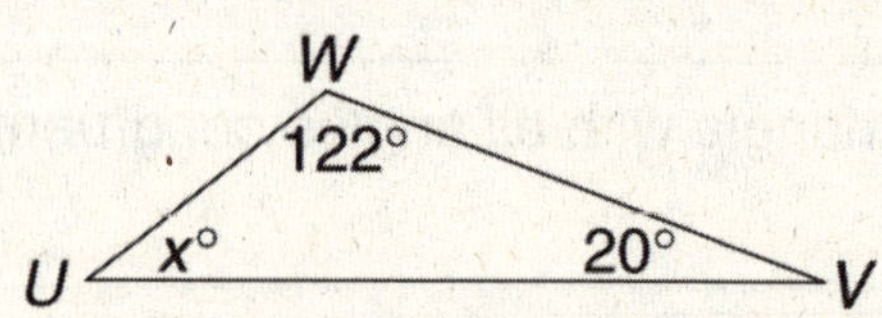

$m\angle U + m\angle V + m\angle W = 180°$

____

____

$x°$ = _______

Name ______________________ Date ____________ Class ____________

LESSON 7-3

Challenge

Change a This into a That

A **geometric dissection** involves cutting a figure into pieces that can then be rearranged to form another figure.

Trace each figure. Cut up the figure you have traced and rearrange the numbered pieces to form the indicated figure. Sketch your solution.

1. Rearrange the pieces of the equilateral triangle to form a square.

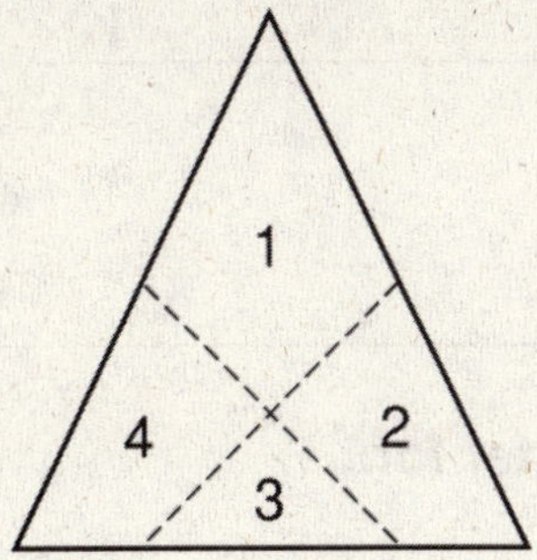

2. Rearrange the pieces of the star to form an equilateral triangle.

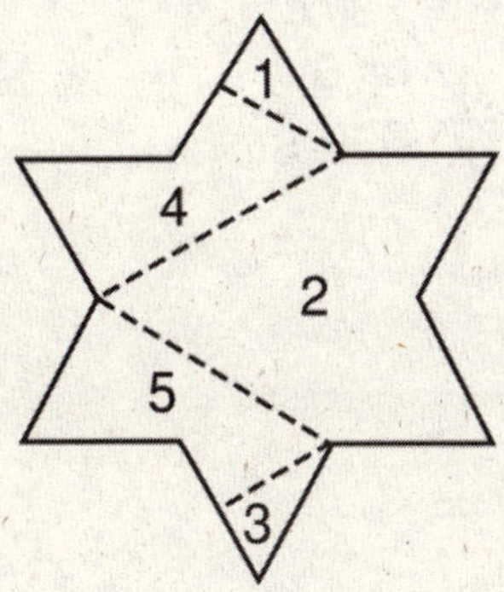

3. Rearrange the pieces of the cross to form an equilateral triangle.

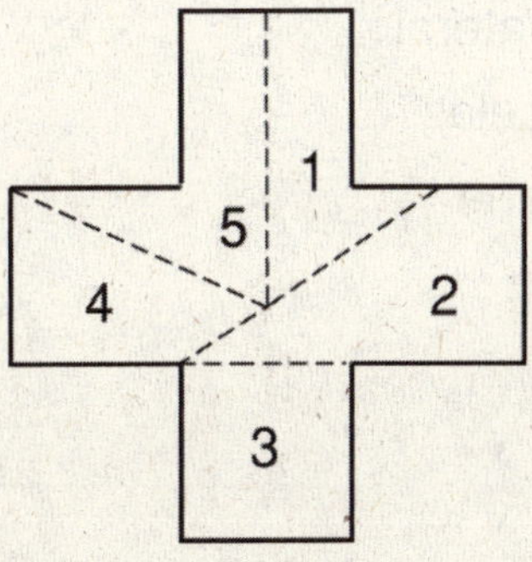

Name ______________________ Date ______________ Class ______________

LESSON 7-3

Problem Solving

Triangles

The American flag must be folded according to certain rules that result in the flag being folded into the shape of a triangle. The figure shows a frame designed to hold an American flag.

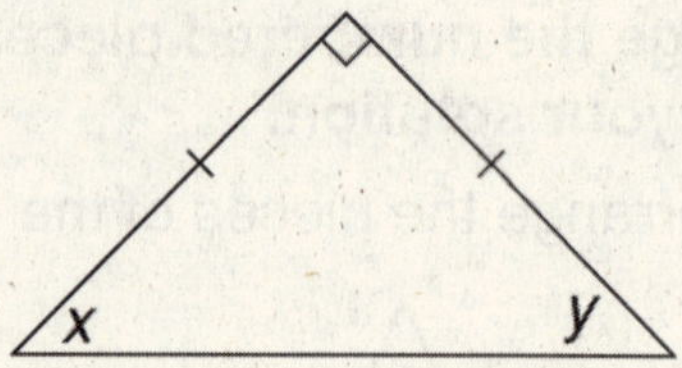

1. Is the triangle acute, right, or obtuse?

____

2. Is the triangle equilateral, isosceles, or scalene?

____

3. Find $x°$.

____

4. Find $y°$.

____

The figure shows a map of three streets. Choose the letter for the best answer.

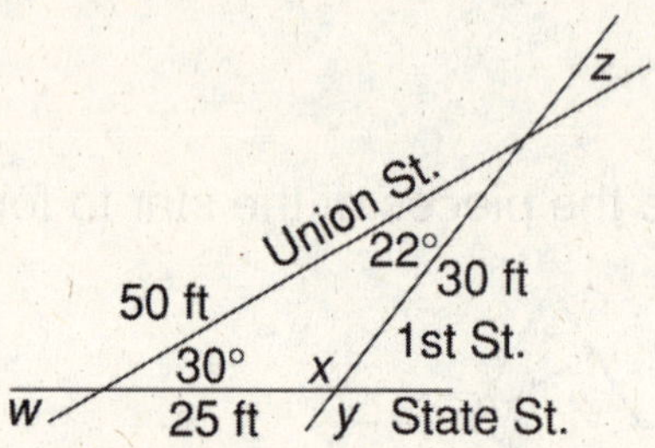

5. Find $x°$.

A 22° C 30°
B 128° D 68°

6. Find $w°$.

F 22° H 30°
G 128° J 52°

7. Find $y°$.

A 22°
B 30°
C 128°
D 143°

8. Find $z°$.

F 22°
G 30°
H 128°
J 143°

9. Which word best describes the triangle formed by the streets?

A acute
B right
C obtuse
D equilateral

10. Which word best describes the triangle formed by the streets?

F equilateral
G isosceles
H scalene
J acute

Name ______________________ Date ____________ Class ____________

LESSON 7-3

Reading Strategies

Graphic Organizer

A triangle can be classified by the measurement of one or more of its angles.

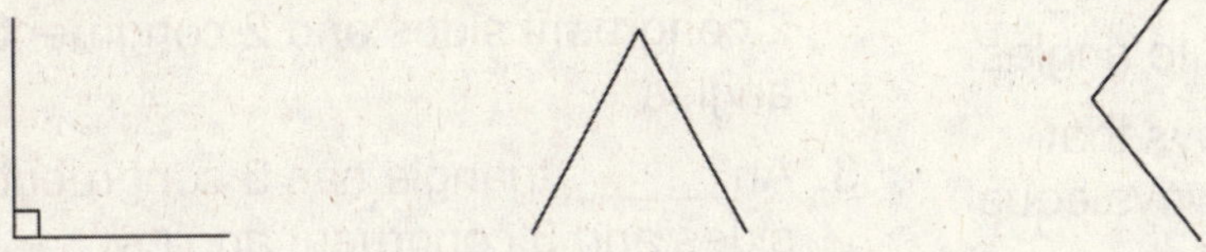

Right angle = 90° Acute angle < 90° Obtuse angle > 90°

A triangle can also be classified by the length of its sides. This chart will help you compare triangles by angles and sides.

Triangles Named by Their Measure

Acute Triangle
All three angles measure less than 90°.

Right Triangle
One angle measures 90°.

Obtuse Triangle
One angle measures more than 90°.

Triangles Named by Length of Sides

Equilateral Triangle
All three sides are equal.

Isosceles Triangle
Two sides are equal.

Scalene Triangle
None of the sides are equal.

Write two of the following terms to describe each triangle: acute, right, obtuse, equilateral, isosceles, scalene.

1.

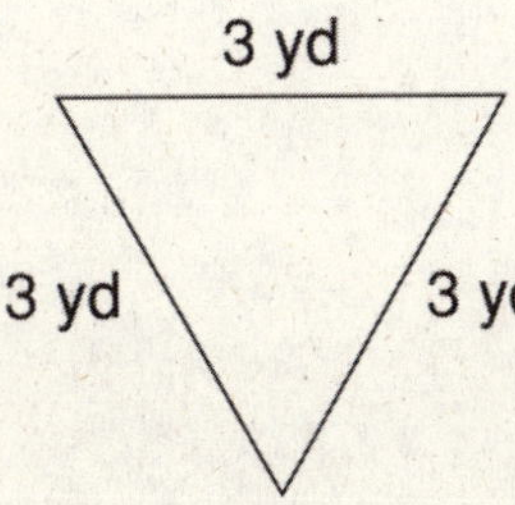

____

2.

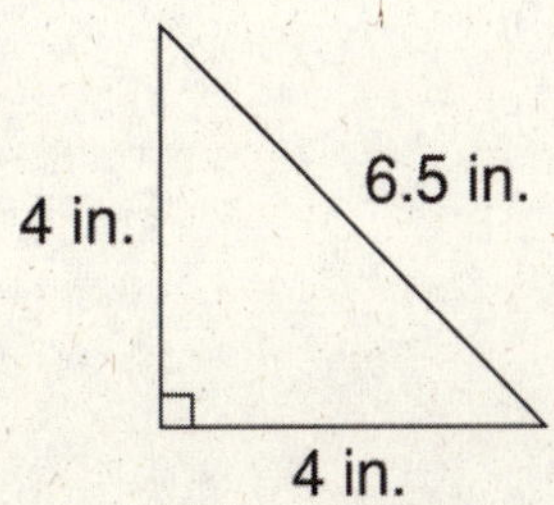

____

Name ______________________ Date ______________ Class ______________

LESSON 7-3

Puzzles, Twisters & Teasers

Try Some Triangles

Across

1. A ______ triangle has one 90° angle.
5. An ______ triangle has 3 acute angles.
6. The Triangle Sum ______ says that the 3 angles in a triangle always equal 180°.
7. A ______ triangle has no congruent sides and no congruent angles.

Down

2. An ______ triangle has at least 2 congruent sides and 2 congruent angles.
3. An ______ triangle has 3 congruent sides and 3 congruent angles.
4. An ______ triangle has 1 obtuse angle.

Name ______________________ Date ______________ Class ______________

LESSON 7-4

Practice A

Polygons

Name each polygon.

1. 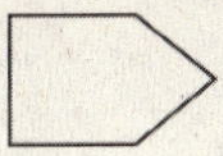

____

2. 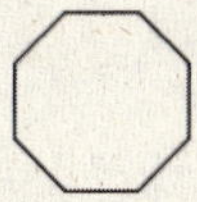

____

3. 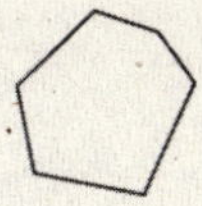

____

Find the sum of the angle measures in each figure.

4. 

____

5. 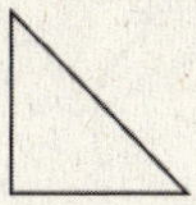

____

6. 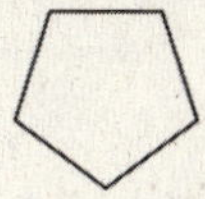

____

7. 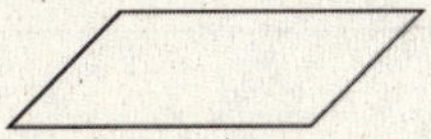

____

8. 

____

9. 

____

Find the angle measures in each regular polygon.

10. 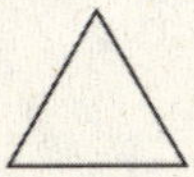

____

11. 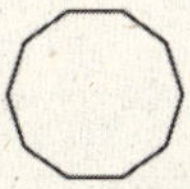

____

12. 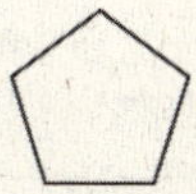

____

13. 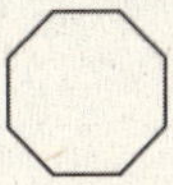

____

14. 

____

15. 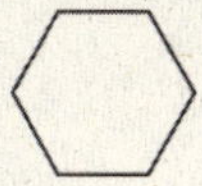

____

Give all the names that apply to each figure.

16. 

____

____

17. 

____

____

18. 

____

____

Name ____________ Date ____________ Class ____________

LESSON 7-4

Practice B

Polygons

Find the sum of the angle measures in each figure.

1. 

2. 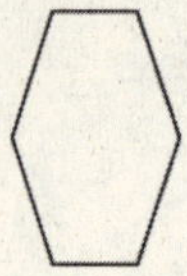

3. 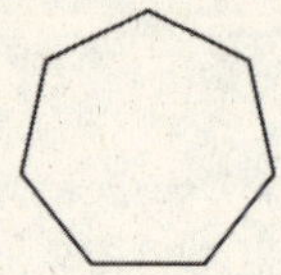

4. 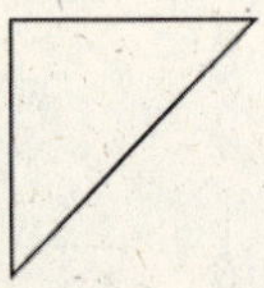

5. 

6. 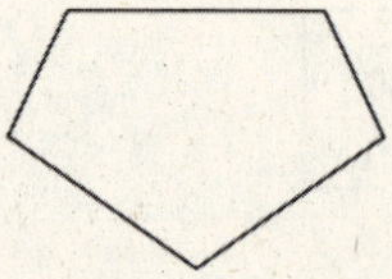

Find the angle measures in each regular polygon.

7. 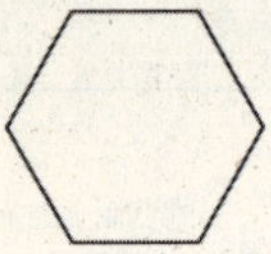

8. 

9. 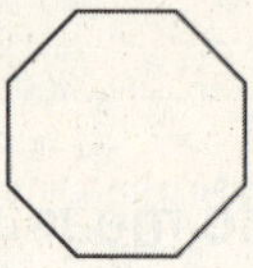

10. 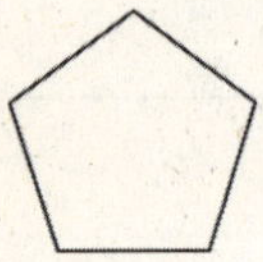

11. 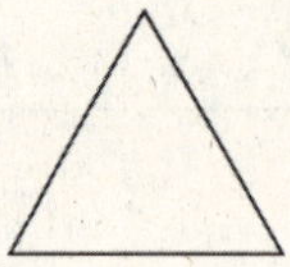

12. 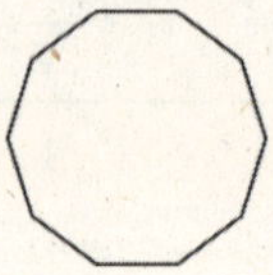

Give all the names that apply to each figure.

13. 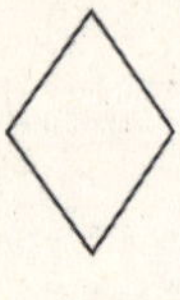

14. 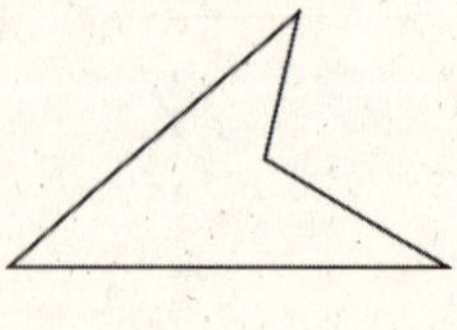

15. 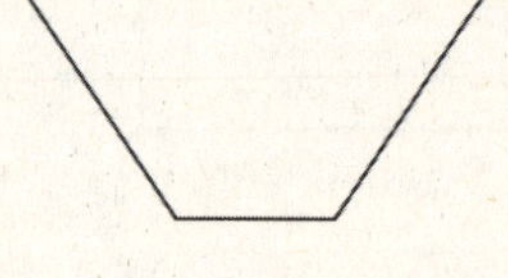

Name ______________________ Date ______________ Class ______________

LESSON 7-4

Practice C

Polygons

Find the sum of the angle measures in each regular polygon. Then, find the measure of each angle.

1. 24-gon
2. 16-gon
3. 36-gon

Find the value of each variable.

4.

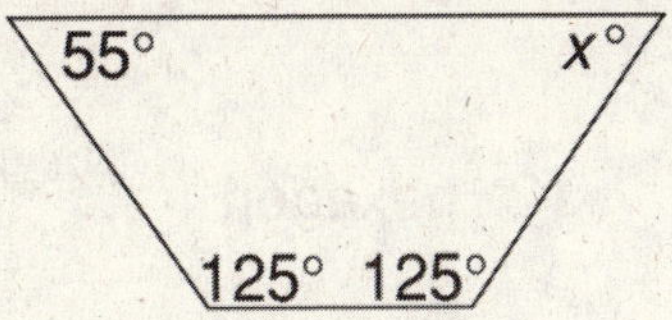

5.

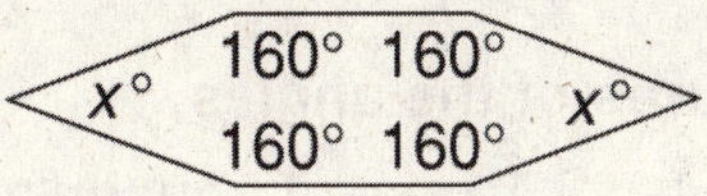

6.

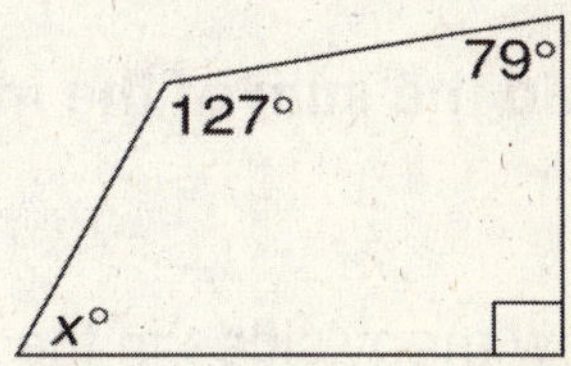

The sum of the angle measures of a polygon is given. Name the polygon.

7. 1080°
8. 1260°
9. 900°
10. 1440°

Graph the given vertices on a coordinate plane. Connect the points to draw a polygon and classify it by the number of its sides.

11. (1, 5), (4, 2), (4, –2), (1, –5), (–3, –5), (–5, –2), (–5, 2), (–3, 5)

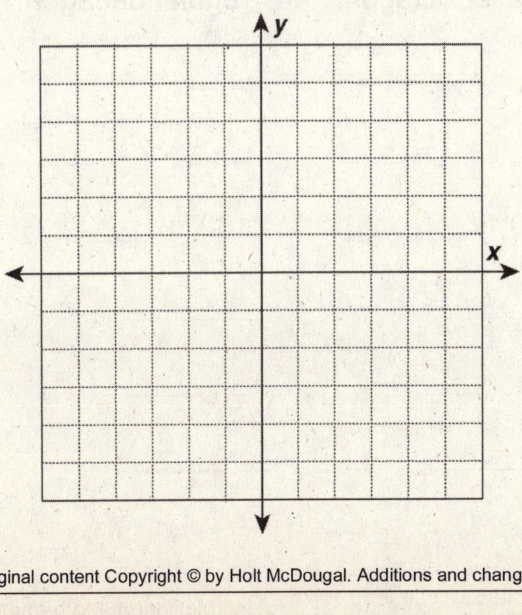

12. (0, –1), (–1, 3), (2, 5), (5, 3), (4, –1)

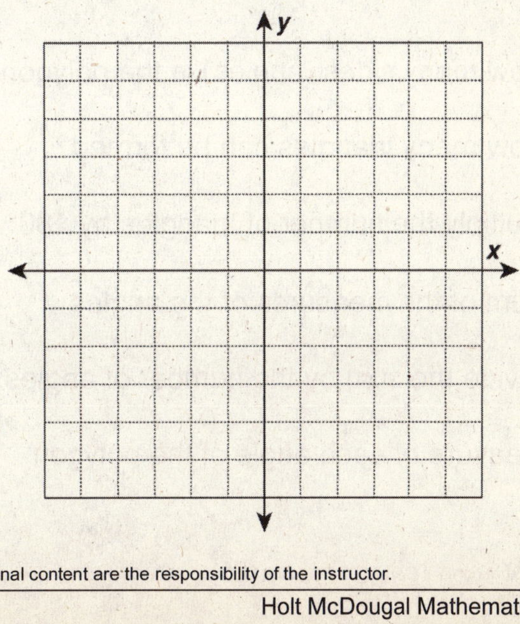

Name ______________________ Date ______________ Class ______________

LESSON 7-4

Review for Mastery

Polygons

A polygon of n sides (an ***n*-gon**) can be divided into $(n-2)$ triangles
The sum of the angle measures of an n-gon $= (n-2)180°$.

A polygon of 5 sides (pentagon) can be divided into 3 triangles.

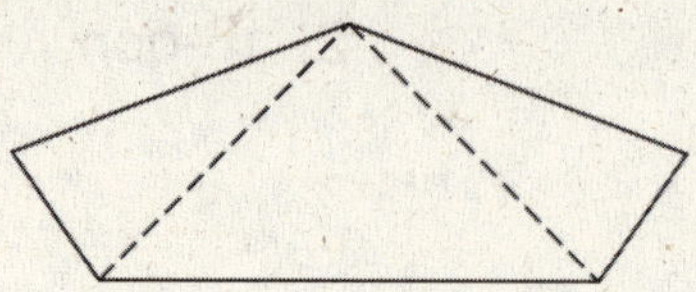

Sum of angle measures of pentagon
$= (n-2)\ 180°$
$= (5-2)\ 180° = (3)180° = 540°$

Find the sum of the measures of the angles.

	1. quadrilateral	2. hexagon
How many sides in the polygon?	____	____
How many triangles can be formed?	____ − 2 = ____	____ − 2 = ____
Multiply the number of triangles by 180°.	180° × ____	180° × ____
Sum of the measures of the angles	________	________

In a **regular polygon**, all sides and all angles are congruent.

The measure of each angle of a regular polygon $= \dfrac{\text{sum of the angles}}{\text{number of sides}}$

The measure of each angle of a regular pentagon $= \dfrac{(5-2)180°}{5} = \dfrac{(3)180°}{5} = \dfrac{540°}{5} = 108°$

Find the measure of each angle.

	3. regular octagon	4. regular decagon
How many sides (angles) in the polygon?	____	____
How many triangles can be formed?	____ − 2 = ____	____ − 2 = ____
Multiply the number of triangles by 180°.	180° × ____	180° × ____
Sum of the measures of the angles	________	________
Divide the sum by the number of angles.	________	________
Measure of each angle of the polygon	________	________

Name ____________________ Date ______________ Class ______________

LESSON 7-4

Review for Mastery

Polygons (continued)

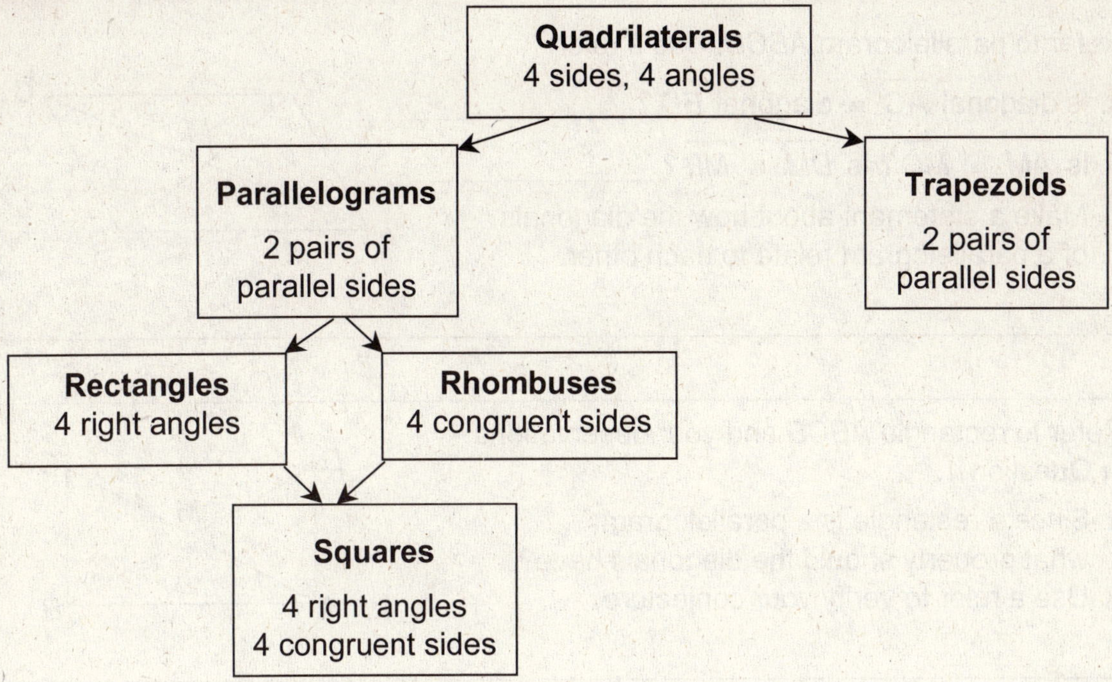

A figure on a lower branch of the tree has the properties of the figures above it.

All rectangles are parallelograms.
But not all parallelograms are rectangles.

Write all the names that apply to each figure.

5. 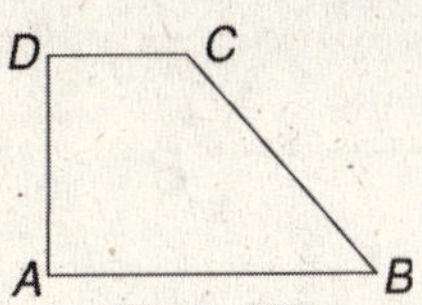

$\overline{AB} \parallel \overline{DC}$

four-sided polygon ______________

1 pair of parallel sides ______________

6. 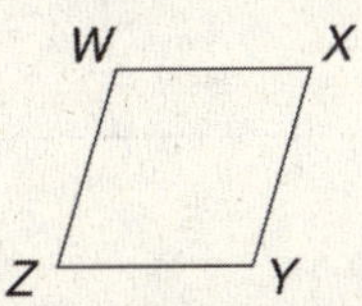

$\overline{WX} \parallel \overline{ZY}$, $\overline{WZ} \parallel \overline{XY}$
$\overline{WX} \cong \overline{XY} \cong \overline{ZY} \cong \overline{WZ}$

four-sided polygon ______________

2 pairs of parallel sides ______________

4 congruent sides ______________

7. 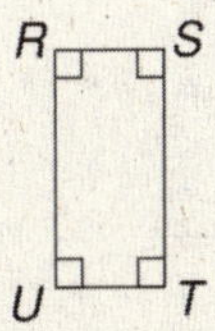

$\overline{RS} \parallel \overline{UT}$, $\overline{RU} \parallel \overline{ST}$

four-sided polygon ______________

2 pairs of parallel sides ______________

4 right angles ______________

Name ______________________ Date ______________ Class ______________

LESSON 7-4

Challenge

Slanted View

1. Refer to parallelogram *ABCD*. Use a ruler.

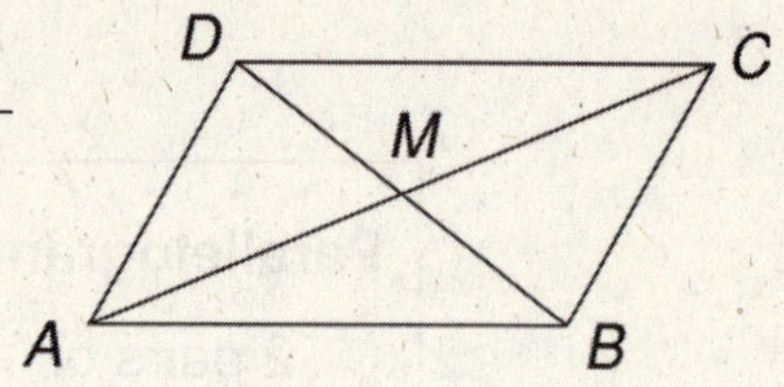

   a. Is diagonal $\overline{AC} \cong$ diagonal $\overline{BD}$? ______________

   b. Is $\overline{AM} \cong \overline{MC}$? Is $\overline{DM} \cong \overline{MB}$? ______________

   Make a statement about how the diagonals of a parallelogram relate to each other.

   ____

   ____

2. Refer to rectangle *ABCD* and your observations in Question 1.

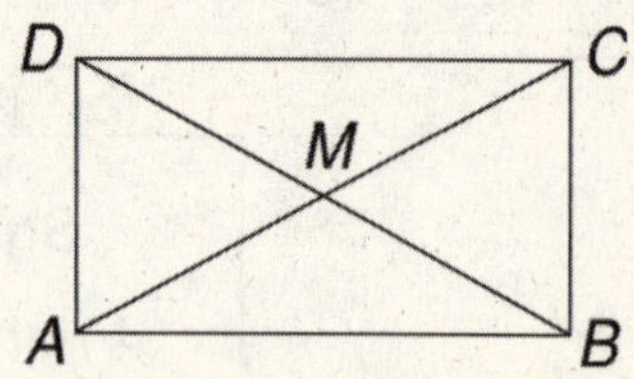

   a. Since a rectangle is a parallelogram, what property should the diagonals have? Use a ruler to verify your conjecture.

   ____

   b. What additional property do the diagonals of a rectangle have?

   ____

3. Refer to rhombus *ABCD*.

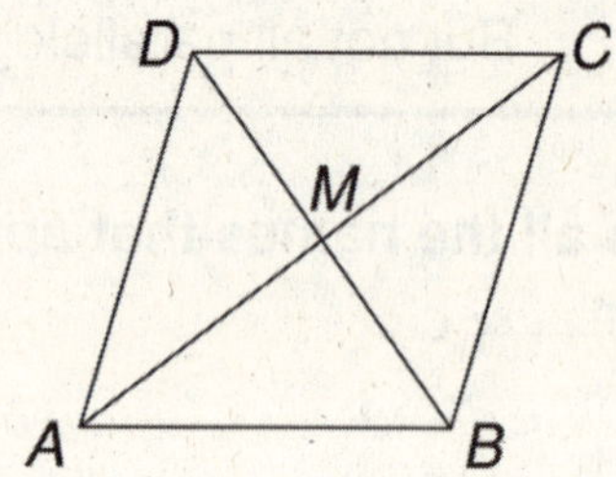

   a. Since a rhombus is a parallelogram, what property should the diagonals have? Use a ruler to verify your conjecture.

   ____

   b. Are the diagonals congruent? ______________

   c. Measure the angles with vertex *M*. What additional property do the diagonals of a rhombus have?

   ____

   d. Measure the angles at each vertex of the rhombus. What additional property do the diagonals of a rhombus have?

   ____

4. Make a conjecture about the properties of the diagonals of a square. Draw a square and verify your conjectures with a ruler and protractor.

   ____

   ____

Name ______________________ Date ______________ Class ______________

LESSON 7-4

Problem Solving

Polygons

The figure shows how the glass for a window will be cut from a square piece. Cuts will be made along, $\overline{CE}$, $\overline{FH}$, $\overline{IK}$ and $\overline{LB}$.

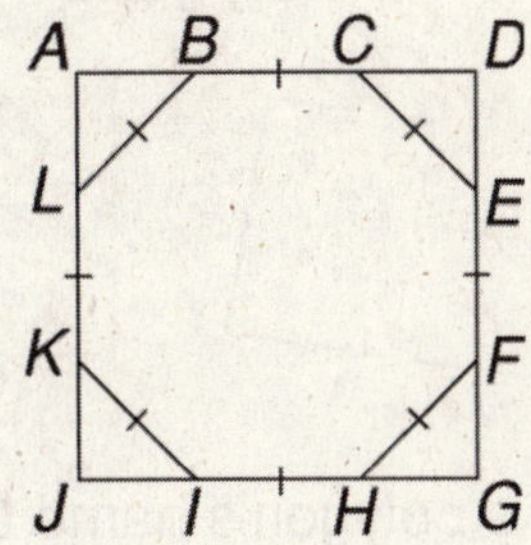

1. What shape is the window?

____

2. What is the sum of the angle measures of the window?

____

3. What is the measure of each angle of the window?

____

4. Based on the angles, what kind of triangle is ΔCDE?

____

5. Based on the sides, what kind of triangle is ΔCDE?

____

The figure shows how parallel cuts will be made along $\overline{AD}$, and $\overline{BC}$. $\overline{AB}$ and $\overline{CD}$ are parallel. Choose the letter for the best answer.

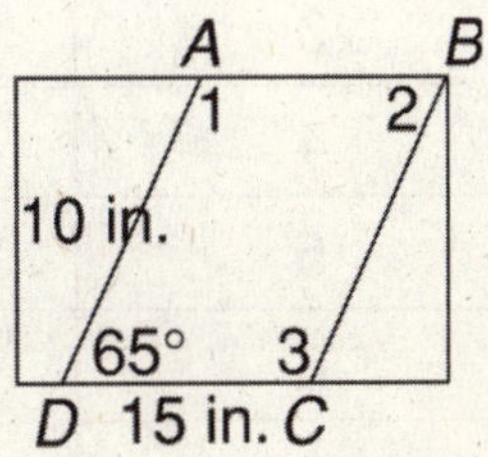

6. Which word correctly describes figure *ABCD* after the cuts are made?
   - A triangle
   - B quadrilateral
   - C pentagon
   - D rhombus

7. Which word correctly describes figure *ABCD* after the cuts are made?
   - A parallelogram
   - B trapezoid
   - C rectangle
   - D hexagon

8. Find the measure of $\angle 1$.
   - A 45°
   - B 65°
   - C 90°
   - D 115°

9. Find the measure of $\angle 2$.
   - F 45°
   - G 90°
   - H 65°
   - J 115°

10. Find the measure of $\angle 3$.
    - A 45°
    - B 90°
    - C 65°
    - D 115°

Name ______________________ Date ______________ Class ______________

LESSON 7-4

Reading Strategies

Reading a Chart

The prefix *poly-* means "many." The word polygon means "many-sided figure." A polygon is a closed, plane figure named by the number of its sides.

Polygons

Not Polygons

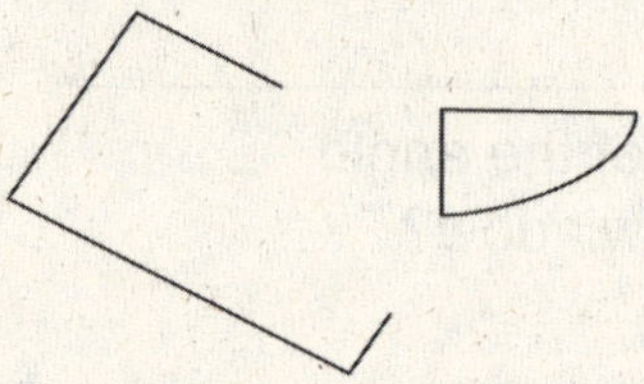

The prefix of a polygon's name tells you the number of its sides and angles.

- *Tri-* means "three" ⟶ A **tri**cycle has 3 wheels.
  ⟶ A **tri**angle has 3 sides and 3 angles.
- *Quad-* means "four" ⟶ A **quad**ruped is a 4-legged animal.
  ⟶ A **quad**rilateral has 4 sides and 4 angles.

This chart helps you organize polygons by their sides and angles.

Polygon	Number of Sides	Number of Angles
Triangle	3	3
Quadrilateral	4	4
Pentagon	5	5
Hexagon	6	6
Octagon	8	8

Write the name of each polygon. Use the chart to help you.

1.

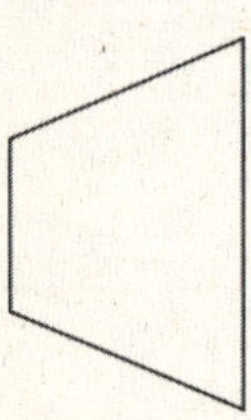

____

2.

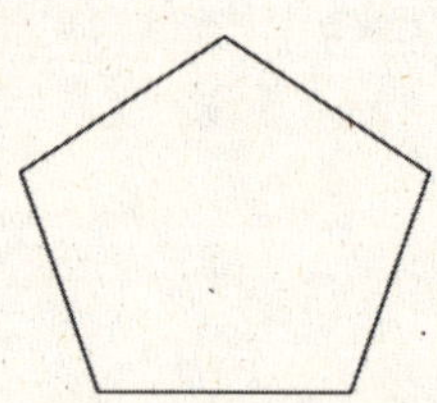

____

3.

____

4.

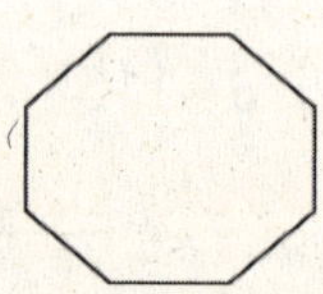

____

Name ____________________ Date ____________ Class ____________

LESSON 7-4

Puzzles, Twisters & Teasers

What Side Are You On?!

Name each figure, one letter on each space. Each answer has one or two boxed letters. Unscramble the boxed letters to solve the riddle.

1. 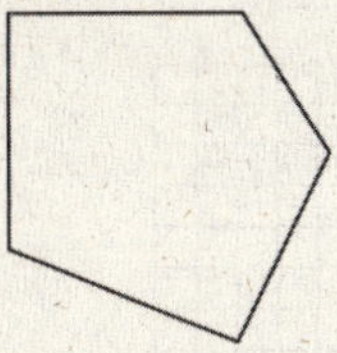

___ ___ ___ ___ [A] ___ ___ ___

2. 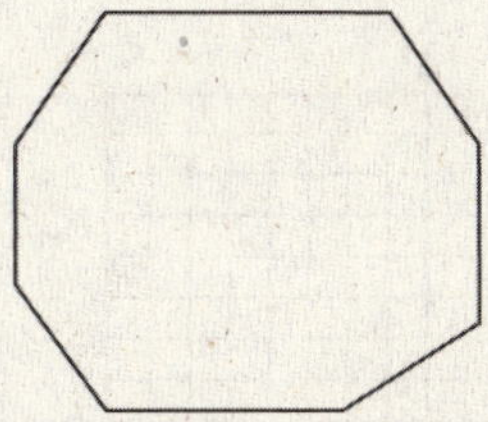

___ ___ ___ [A] ___ ___ ___

3. 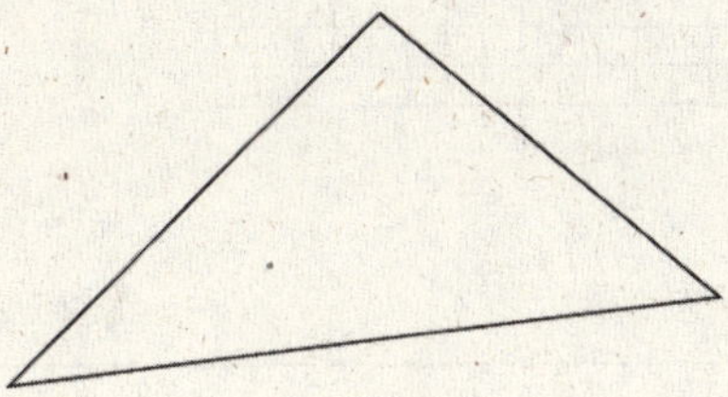

___ ___ [I] ___ ___ ___ ___ ___

4. 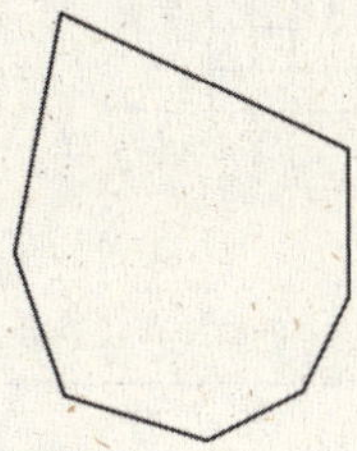

[H] ___ ___ ___ ___ ___ ___

5. 

___ ___ ___ [D] ___ ___ ___ ___ ___ ___ [R] ___ ___

6. 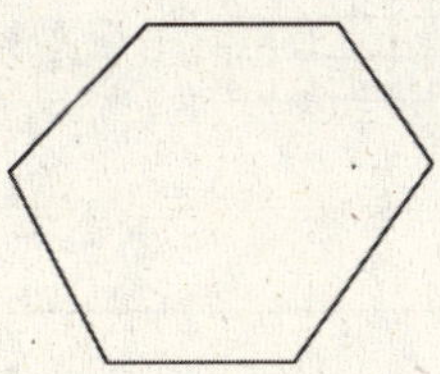

___ [E] ___ ___ ___ ___ ___

Why was the computer tired after its long trip?

It was ___ ___ ___ R ___ D ___ ___ V ___ .

Name ______________________ Date ______________ Class ______________

LESSON 7-5

Practice A

Coordinate Geometry

Graph the polygons with the given vertices. Give the most specific name for each polygon.

1. $A(0, 6)$, $B(-3, 0)$, $C(3, 0)$

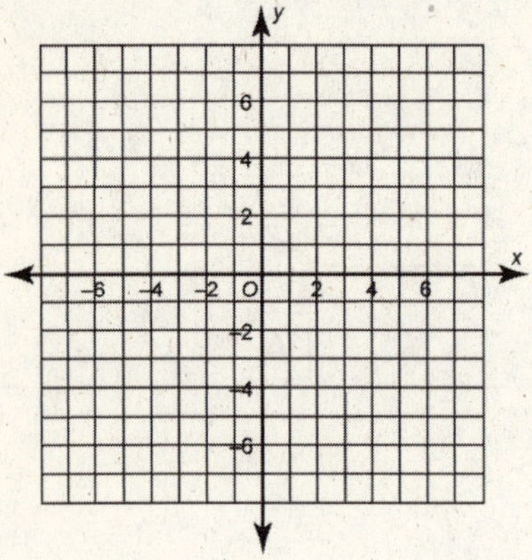

____

2. $W(-2, 4)$, $X(-2, -3)$, $Y(5, -3)$, $Z(5, 4)$

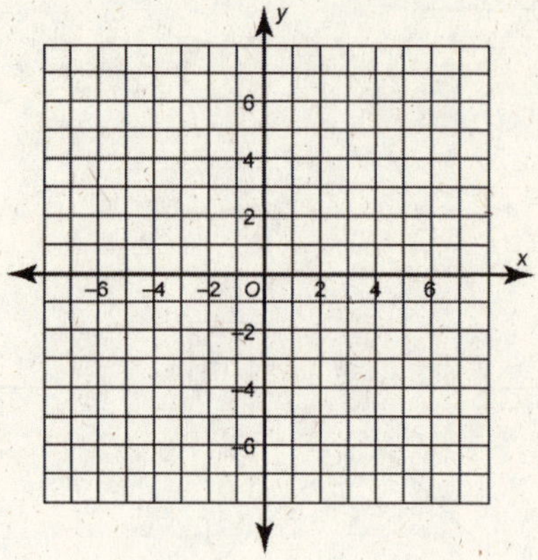

____

Find the coordinates of each missing vertex.

3. Triangle *MNP* has a right angle at *M* and *MP* = 4.

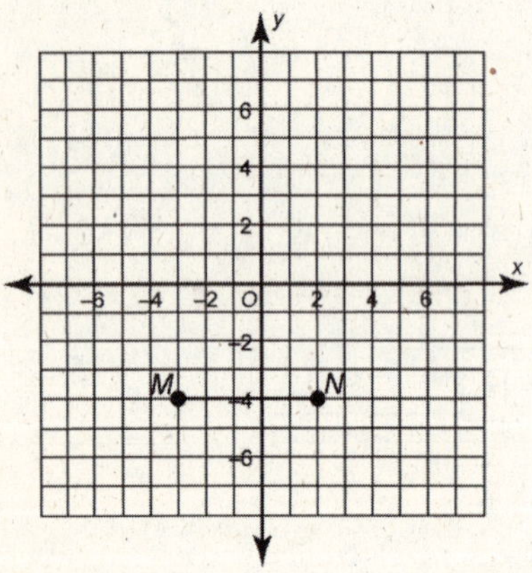

____

4. Quadrilateral *FGHK* is a rectangle.

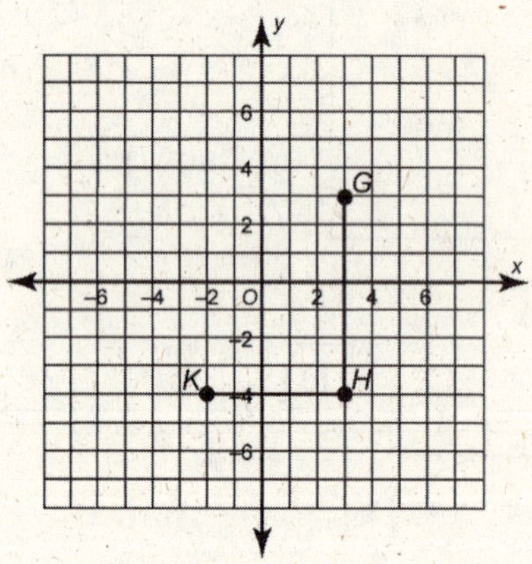

____

Find the coordinates of the midpoint of each segment.

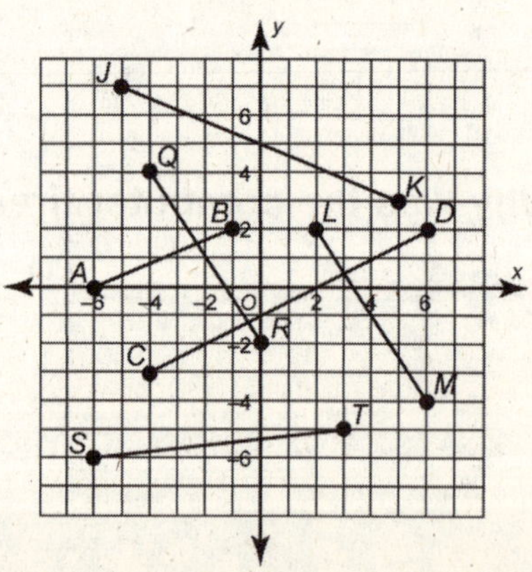

5. $\overline{AB}$ ______________

6. $\overline{CD}$ ______________

7. $\overline{ST}$ ______________

8. $\overline{QR}$ ______________

9. $\overline{JK}$ ______________

10. $\overline{LM}$ ______________

Name ______________________ Date ______________ Class ______________

LESSON 7-5

Practice B

Coordinate Geometry

Graph the polygons with the given vertices. Give the most specific name for each polygon.

1. $S(-3, -3)$, $T(-3, 4)$, $U(4, -3)$

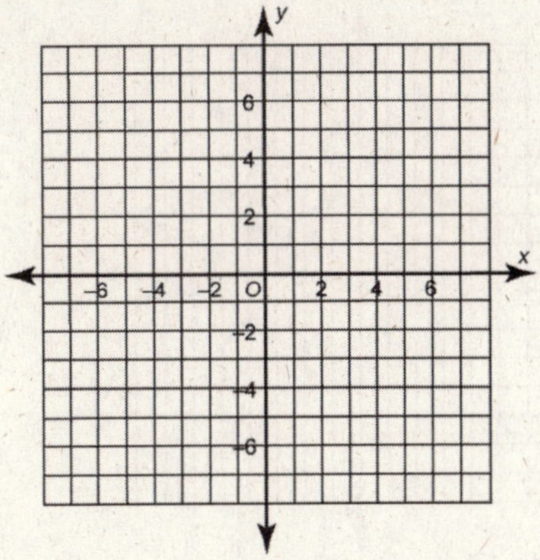

____

2. $M(-1, 2)$, $N(2, 0)$, $Q(2, -4)$, $P(-1, -2)$

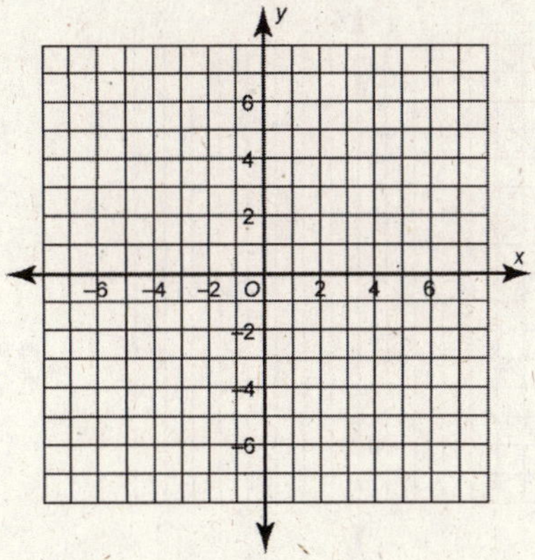

____

Find the coordinates of each missing vertex.

3. Triangle *WUV* has a right angle at *W* and $WU = 5$.

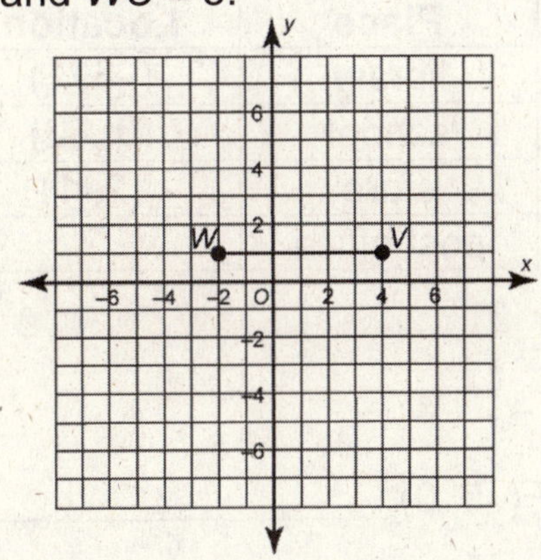

____

4. Quadrilateral *JKLM* is a square.

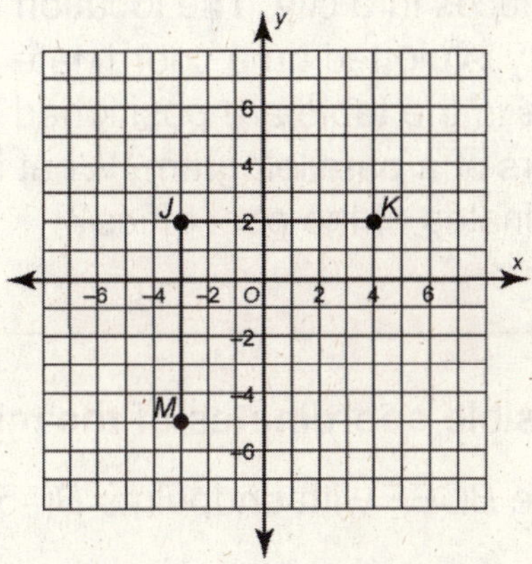

____

Find the coordinates of the midpoint of each segment.

5. $\overline{XY}$ with endpoints $X(1, 2)$ and $Y(4, 5)$ ______________________

6. $\overline{ST}$ with endpoints $S(-8, 6)$ and $T(2, 4)$ ______________________

7. $\overline{FG}$ with endpoints $F(3, 9)$ and $G(-5, 6)$ ______________________

8. $\overline{LP}$ with endpoints $L(6, 0)$ and $P(-3, -4)$ ______________________

9. $\overline{EC}$ with endpoints $E(-2, -8)$ and $C(-1, -7)$ ______________________

Name ______________________ Date ______________ Class ______________

LESSON 7-5

Practice C

Coordinate Geometry

Graph the polygons with the given vertices. Give the most specific name for each polygon.

1. $W(-2, 4)$, $X(5, 2)$, $Y(5, -1)$

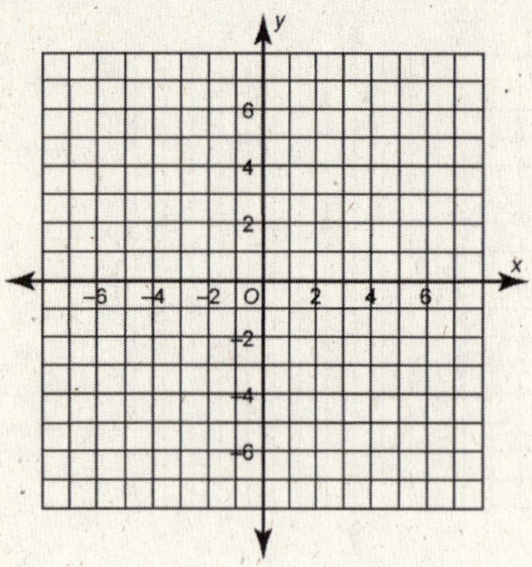

____

2. $J(0, 1)$, $K(4, 1)$, $L(4, -3)$, $M(-3, -3)$

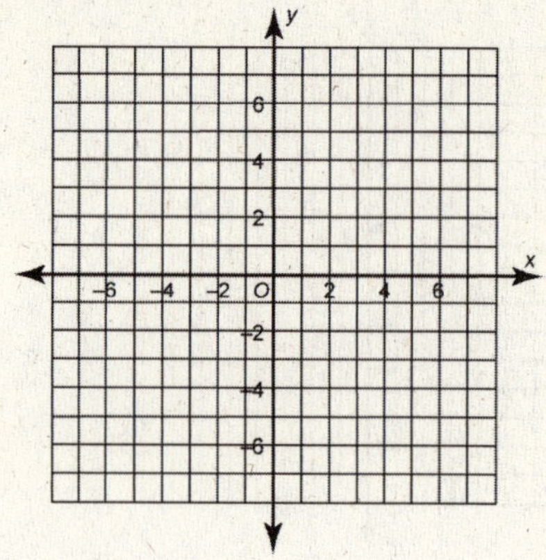

____

Find the missing coordinates.

3. For a school project, Mark is designing a map of four places in a city. The location of each place in the city is plotted on a coordinate grid. The four places in the table are positioned at the four corners of a parallelogram. What should be the coordinates of the post office?

Map Locations	
Place	**Location**
library	(–3, 1)
school	(0, –2)
store	(2, 1)
post office	?

____

Find possible coordinates of the missing vertex.

4. square *BCEF* with endpoints $B(-5, 7)$, $C(2, 7)$, and $F(-5, 0)$ ______________

5. isosceles triangle *FGH* with $F(-4, 4)$, $H(3, -3)$, and right angle *G* ______________

6. rectangle *WXYZ* with endpoints $W(-1, 5)$, $Y(-1, -2)$, and $Z(7, -2)$ ______________

7. parallelogram *ABCD* with endpoints $B(0, 3)$, $C(4, 3)$, and $D(2, -1)$ ______________

Find the coordinates of the midpoint of each segment.

8. $\overline{CD}$ with endpoints $C(-6, 0)$ and $D(7, -1)$ ______________

9. $\overline{MN}$ with endpoints $M(0, -8)$ and $N(-2, -7)$ ______________

10. $\overline{BL}$ with endpoints $B(8, -3)$ and $L(0, -4)$ ______________

Name ______________________ Date ______________ Class ______________

LESSON 7-5

Review for Mastery

Coordinate Geometry

When you have a polygon on a coordinate grid, look for parallel lines, perpendicular lines, and right angles to help you classify the polygon.

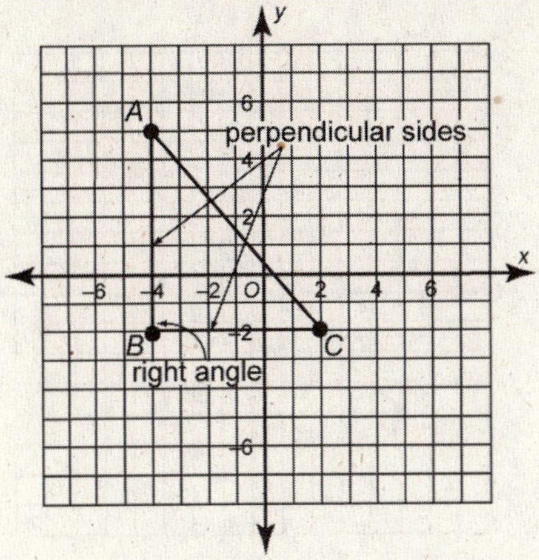

This is a right triangle. It has one right angle. It has a set of perpendicular sides

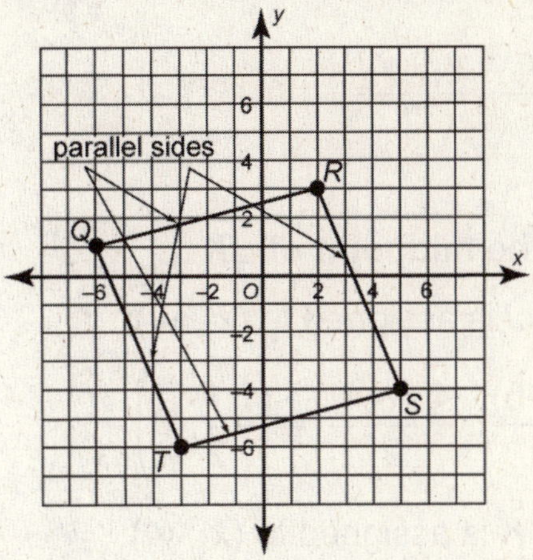

This is a parallelogram. It has two sets of parallel lines.

Graph each polygon with the given vertices. Tell if the polygon has any parallel lines, perpendicular lines, or right angles. Give the most specific name for the polygon.

1. $C(-4, 7)$, $D(6, 7)$, $E(6, -3)$, $F(-4, 3)$

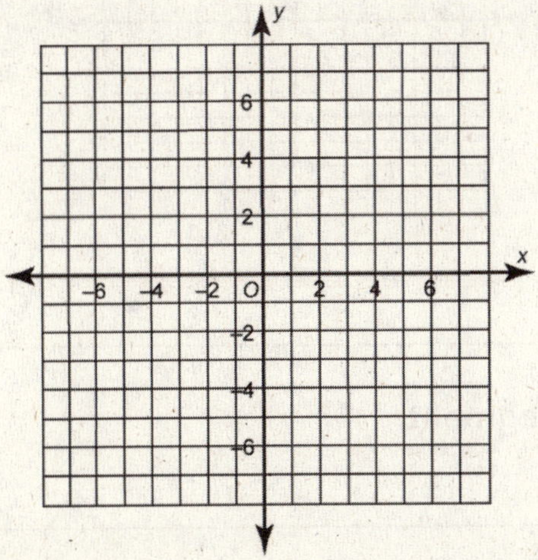

____

____

____

2. $L(-3, 1)$, $M(0, 1)$, $N(1, -4)$, $Q(-5, -4)$

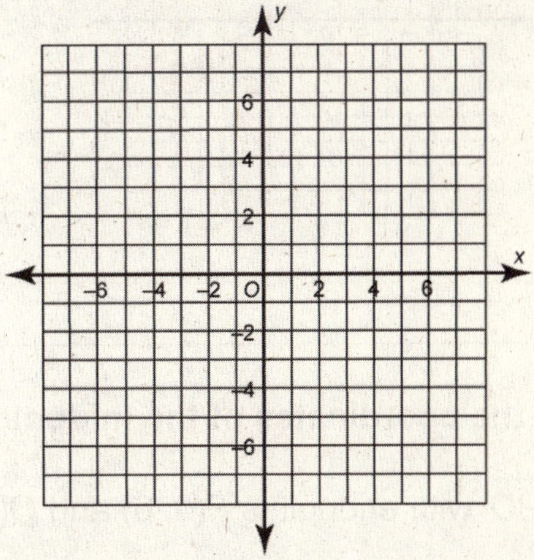

____

____

____

Name ______________________ Date ______________ Class ______________

LESSON 7-5

Review for Mastery

Coordinate Geometry (continued)

The midpoint of a segment is the point that is halfway between the endpoints. Use the formula to find a point. You can assign either endpoint to (x_1, y_1) and get the same midpoint.

$$\left(\frac{x_1 + x_2}{2}, \frac{y_1 + y_2}{2}\right)$$

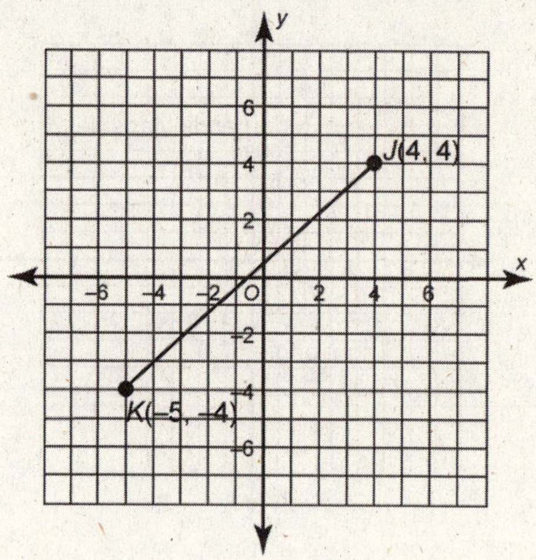

Find the midpoint of $\overline{JK}$.

When *J* is assigned to (x_1, y_1):

$$\left(\frac{4+(-5)}{2}, \frac{4+(-4)}{2}\right) = \left(-\frac{1}{2}, 0\right)$$

When *K* is assigned to (x_1, y_1):

$$\left(\frac{-5+4}{2}, \frac{-4+4}{2}\right) = \left(-\frac{1}{2}, 0\right)$$

Find the midpoint of each segment.

3.

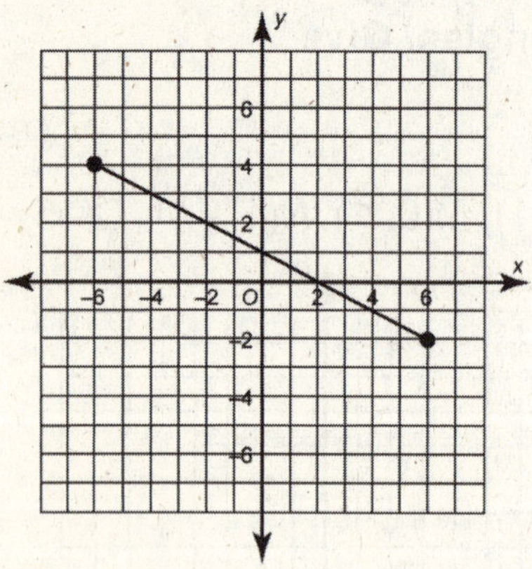

____

4.

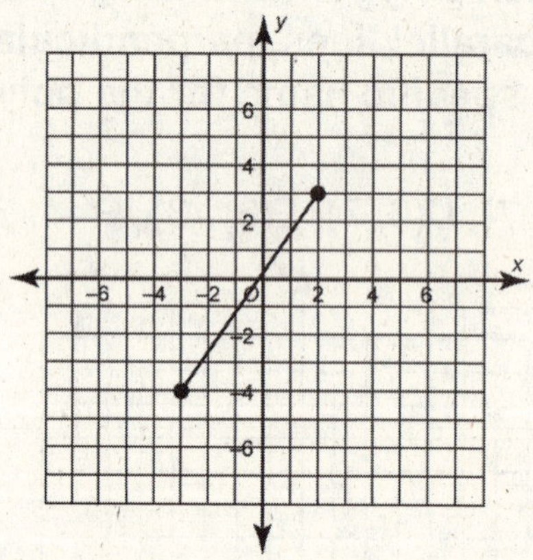

____

Find the coordinates of the midpoint of each segment.

5. $\overline{PQ}$ with endpoints *P*(0, 6) and *Q*(6, 0) ______________________

6. $\overline{VW}$ with endpoints *V*(−2, 6) and W(4, 2) ______________________

7. $\overline{GH}$ with endpoints *G*(−1, 7) and *H*(−3, 4) ______________________

Name ______________________ Date ____________ Class ____________

LESSON 7-5

Challenge

Endpoints and Midpoints

Find the coordinates of each missing endpoint when the coordinates of one endpoint and the coordinates of the midpoint are known.

1. Find the coordinates of *C* if *B* is the midpoint of $\overline{AC}$.

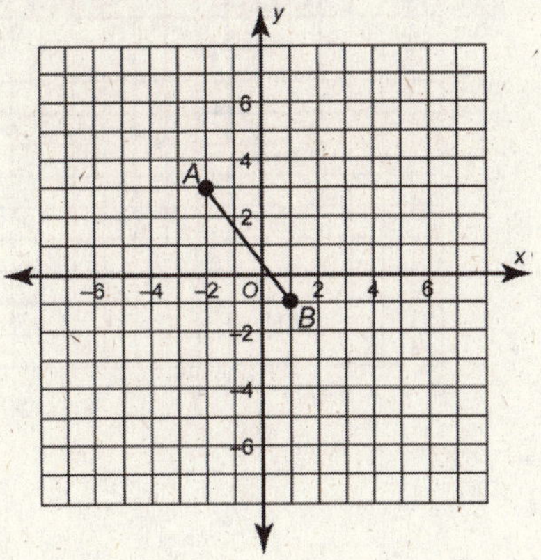

2. Find the coordinates of *X* if *Y* is the midpoint of $\overline{XZ}$.

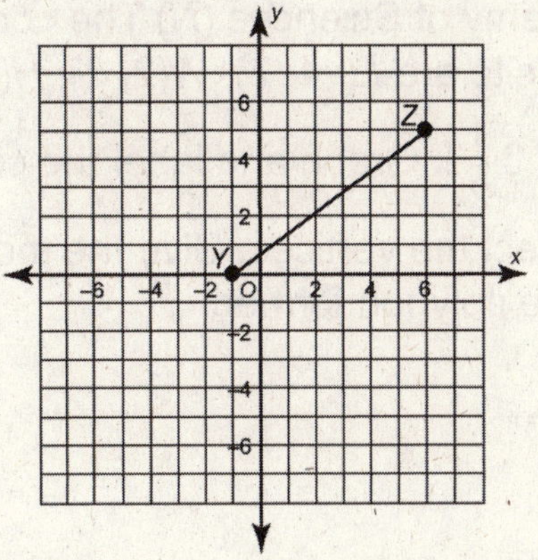

______________________ ______________________

3. Find the coordinates of *M* if *N*(–1, 4) is the midpoint of $\overline{MP}$ and *P* has coordinates (2, –3).

____

4. Find the coordinates of *H* if *G*(9, 0) is the midpoint of $\overline{FH}$ and *H* has coordinates (–4, –5).

____

The Sierpinski Triangle, a famous geometric design, is made when the midpoints of the sides of an equilateral triangle are connected to form additional triangles. This process can be repeated indefinitely, thus creating a self-similar image.

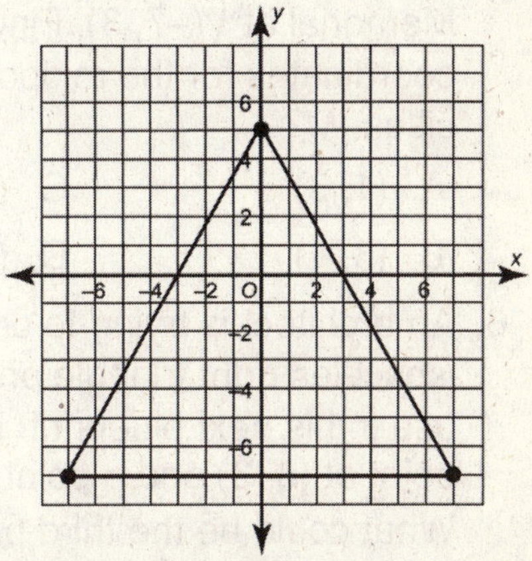

5. What are the midpoints of each side of the triangle at the right? ______________________

6. When the midpoints are connected, how many congruent triangles are formed? ____________

7. When the process is repeated for each new triangle created, how many congruent triangles are then formed? ______________________

Name ______________________ Date ______________ Class ______________

LESSON 7-5

Problem Solving

Coordinate Geometry

Nguyen and his family went to Washington D.C. Nguyen graphed the sites he and his family visited on a coordinate grid.

1. On Saturday, Nguyen and his family visited the Lincoln Memorial (*L*), the Washington Monument (*W*), the White House (*H*), and the National Academy of Sciences (*N*) The coordinates for each site are *L*(–4, 2), *W*(2, 2), *H*(2, 6) and *N*(–3, $3\frac{1}{2}$). Graph and label the coordinates. Connect the vertices. Give the most specific name for the polygon formed.

   ____

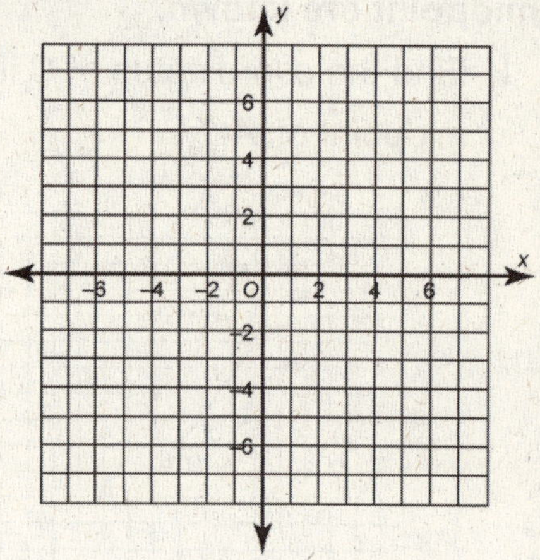

2. The White House, the Washington Memorial, and the Smithsonian form triangle *HWS*. Triangle *HWS* has a right angle at *W*. $\overline{WS}$ is $3\frac{1}{2}$ units. What are possible coordinates of the Smithsonian (*S*)? ____________________

3. The Department of the Interior is located at the midpoint of the White House and the Lincoln Memorial. What are the coordinates where the Department of the Interior is located? ____________________.

Choose the letter for the best answer.

4. A city map shows the State Capital at *C*(5, 3) and a Vietnam Veterans Memorial at *V*(–7, 3). Find the coordinates for the midpoint of the cities, *M*?

   A (1, 3) C (–3, 1)
   B (3, 1) D (–1, 3)

5. The location of four boats are shown on a coordinate grid. The boats are located at *S*(5, 3), *H*(5, 5) *I*(–7, 5), and *P*(–7, 3). What polygon is formed by the ships?

   F kite H square
   G rectangle J trapezoid

6. An architect is trying to draw an isosceles right triangle on a grid to use in his next blueprint. He has a point at (6, 3) and a point at (–2, 3). What could be the third point?

   A (–2, 0) C (2, –2)
   B (6, –5) D (6, 8)

7. Two points on an electrician's diagram are located at (5, 3) and (–7, 3). The scale on the diagram is 1 unit = 0.25 m. What is the actual distance between the two points?

   F 1 m H 3 m
   G 2 m J 4 m

Name ______________________ Date ______________ Class ______________

LESSON 7-5

Reading Strategies

Make Connections

When a figure is on a coordinate plane, find the lengths of vertical and horizontal segments by counting units. Also use the fact that horizontal and vertical lines are perpendicular.

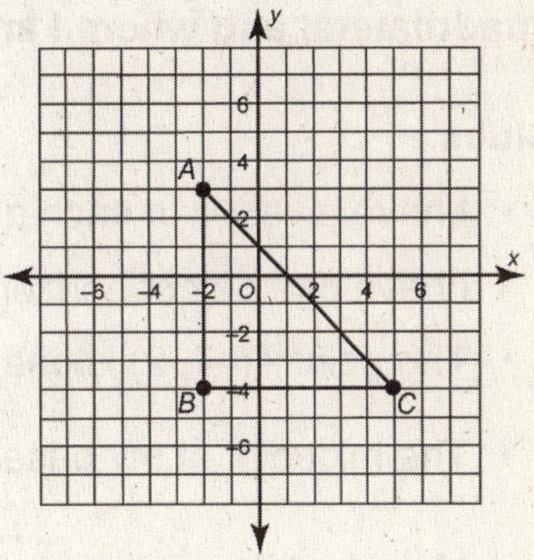

- $\overline{AB}$ is vertical and $\overline{BC}$ is horizontal. So $\overline{AB}$ and $\overline{BC}$ are **perpendicular**.
- $\overline{AB}$ and $\overline{BC}$ are both 7 units long, so they are **congruent**.

A right triangle with two congruent sides is an **isosceles right triangle**.

1. Graph the coordinates (4, 0), (3, 3), (–1, 3) and (0, 0). Connect the vertices.

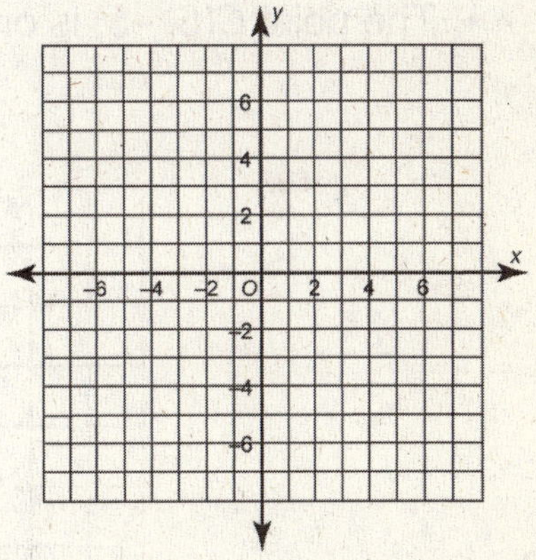

2. How many pairs of parallel sides does the figure have? ______________
3. How many right angles does the figure have? ______________
4. How many pairs of congruent sides does the figure have? ______________
5. What is the **best** name for the figure? ______________

The **midpoint** of a line segment is exactly in the middle between the two endpoints. You can substitute the coordinates of the endpoints into the **midpoint formula** to find the midpoint.

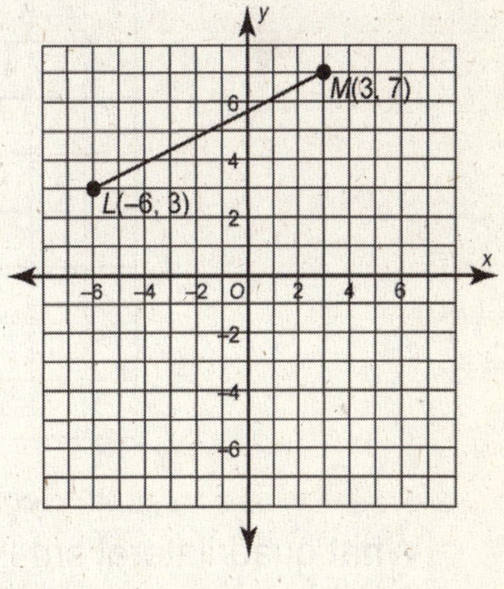

Midpoint formula: $\left(\frac{x_1 + x_2}{2}, \frac{y_1 + y_2}{2}\right)$

Replace (x_1, y_1) with (–6, 3): $\left(\frac{-6 + x_2}{2}, \frac{3 + y_2}{2}\right)$

Then replace (x_2, y_2) with (3, 7): $\left(\frac{-6 + 3}{2}, \frac{3 + 7}{2}\right)$

Simplify: $\left(\frac{-3}{2}, \frac{10}{2}\right) = \left(-1\frac{1}{2}, 5\right)$.

6. Find the midpoint of a segment with endpoints (–5, 7) and (1, 1). ______________
7. Find the midpoint of a segment with endpoints (3, –1) and (6, 4). ______________

Name ______________________ Date ______________ Class ______________

LESSON 7-5

Puzzles, Twisters & Teasers

Shape Sleuth

I am a quadrilateral. Use the clues to find which type of quadrilateral and where I am located on the coordinate grid.

Clues

- I have a vertex in each quadrant.
- I have one pair of horizontal parallel lines.
- The point *A*(–4, 2) is one of my vertices.
- The midpoint of my base, $\overline{CD}$, is (–1, –5).
- My side *AB* is $\frac{2}{3}$ the length of side *CD*.
- The point *C*(5, –5) is one of my vertices.

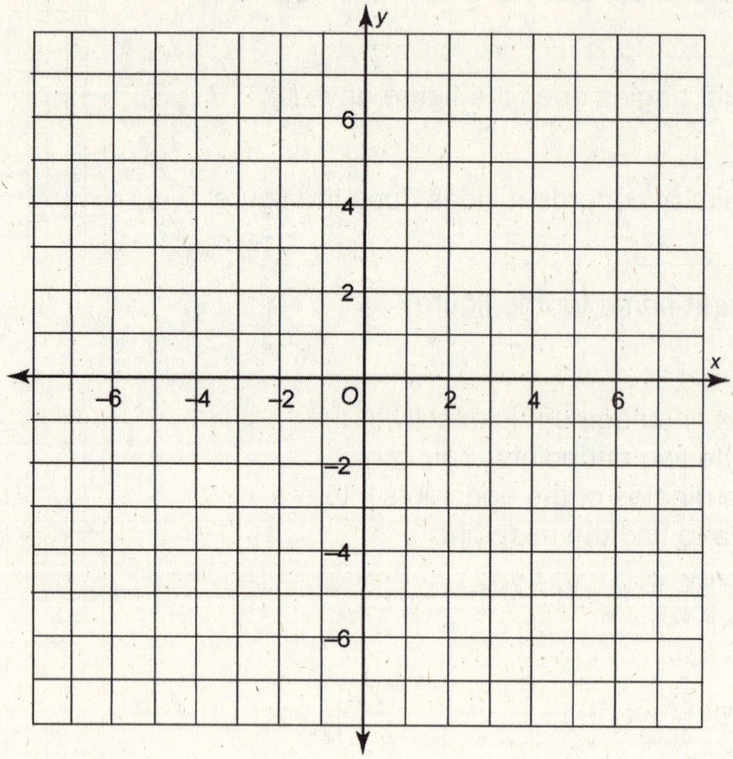

What quadrilateral am I? ______________________

What are the coordinates of my vertices?

A ____________ *B* ____________

C ____________ *D* ____________

CHAPTER 7

Family Letter

7B Patterns in Geometry

Dear Family,

The student will begin to apply geometric properties to solve problems. One such property addresses congruent figures. As the student previously learned, if two polygons are congruent, then all of their corresponding sides and angles are also congruent.

The student will write congruence statements for pairs of polygons by writing the second polygon in order of correspondence, or matching up the corresponding vertices between the two polygons.

Write a congruence statement for this pair of polygons.

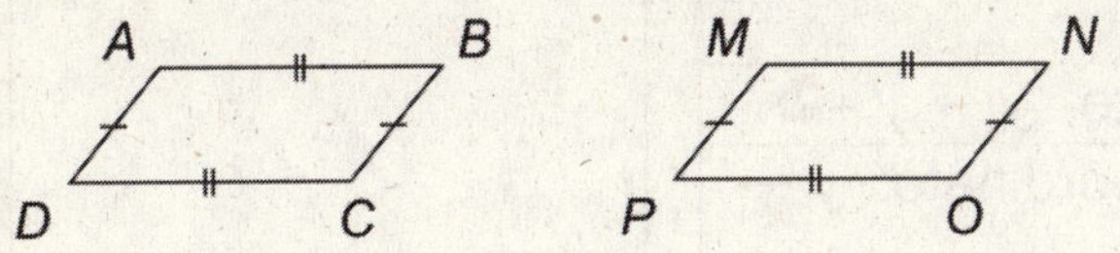

$\angle A \cong \angle M$ so $\angle A$ corresponds to $\angle M$.

$\angle B \cong \angle N$ so $\angle B$ corresponds to $\angle N$.

$\angle C \cong \angle O$ so $\angle C$ corresponds to $\angle O$.

$\angle D \cong \angle P$ so $\angle D$ corresponds to $\angle P$.

The congruence statement is
parallelogram $ABCD \cong$ parallelogram $MNOP$.

Note how the vertices in the first polygon are written in order around the polygon starting at any vertex. The vertices in the second polygon, therefore, have to be written in the same order.

If you know polygons are congruent, you can find an unknown value in the polygon.

In the figure, triangle RST ≅ triangle *UVW*. Find *m*.

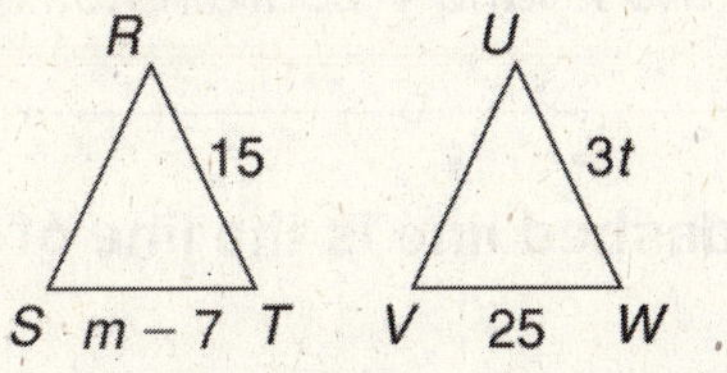

$m - 7 = 25$ $\quad ST \cong VW$

$$\begin{aligned} m - 7 &= 25 \\ +7 \quad &\;\; +7 \\ \hline m &= 32 \end{aligned}$$

Add 7 to both sides

The value of m is 32.

Vocabulary

These are the math words we are learning:

center of rotation the point about which a figure is rotated

congruent figures figures that have the same size and shape

correspondence a way of matching up two sets of objects

image the resulting figure after a translation, rotation, or reflection

line of symmetry the line that creates line symmetry

line symmetry occurs when a line is drawn through a figure and two sides are mirror images of each other

reflection flips a figure across a line to create a mirror image

rotation turns a figure around a point

rotational symmetry occurs when a figure is rotated less than 360° around a central point so it coincides with itself

Family Letter

7B Patterns in Geometry *continued*

Another set of geometric concepts the student will be learning about is **transformations.** Transformations include rotations, reflections, and translations. The result of those transformations are congruent to the original figure.

If you move a figure along a line, it is called a **translation.** If you turn a figure around a point, it is called a **rotation.** If you flip a figure across a line to create a mirror image, it is called a **reflection.**

Translations	
Type	Rule
Move right *a* units	Add *a* to each *x*-coordinate: $x \to x + a$
Move left *a* units	Subtract *a* from each *x*-coordinate: $x \to x - a$
Move up *b* units	Add *b* to each *y*-coordinate: $y \to y + b$
Move down *b* units	Subtract *b* from each *y*-coordinate: $y \to y - b$

Reflections	
Type	Rule
Across the *y*-axis	Multiply each *x*-coordinate by –1: $x \to -x$
Across the *x*-axis	Multiply each *y*-coordinate by –1: $y \to -y$

Rotations	
Type	Rule
180°	Multiply both coordinates by –1: $(x, y) \to (-x, -y)$
90° clockwise	Multiply each *x*-coordinate by –1, then switch the *x*- and *y*-coordinates: $(x, y) \to (y, -x)$
90° counter-clockwise	Multiply each *y*-coordinate by –1, then switch the *x*- and *y*-coordinates: $(x, y) \to (-y, x)$

Complete the figure. The dashed line is the line of symmetry.

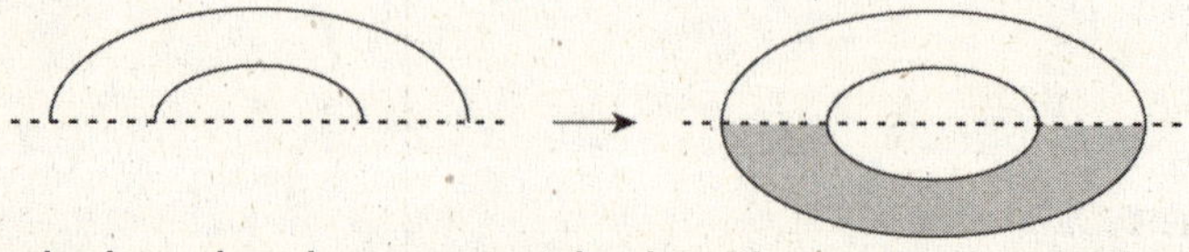

The shaded region is symmetrical to the non-shaded region.

Sincerely,

transformation translating, reflecting, or rotating an object

translation slides a figure along a line without turning

Name ____________________ Date ______________ Class ______________

CHAPTER 7

At-Home Practice

7B Patterns in Geometry

Write a congruence statement for each pair of polygons.

1. 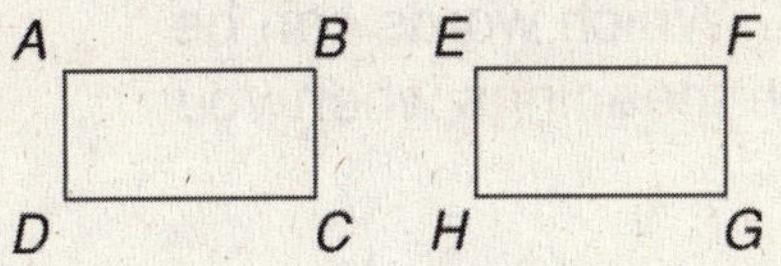

2. 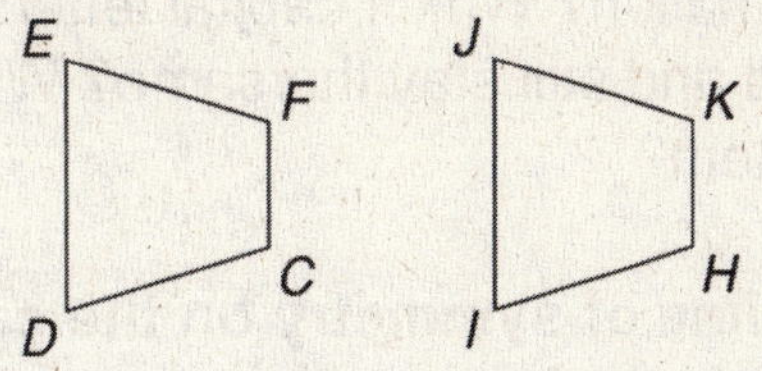

________________________ ________________________

In the figure, triangle *DEF* ≅ triangle *PQR*. Find the value of the variable.

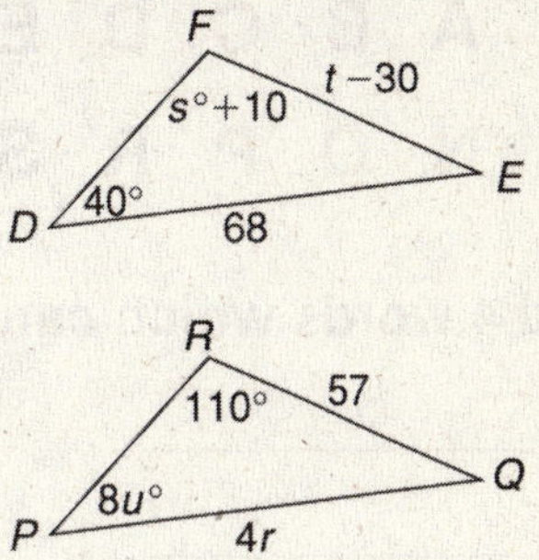

3. Find *r*. ________

4. Find *s*. ________

5. Find *t*. ________

6. Find *u*. ________

Identify the coordinates of each vertex of the image.

7. Translate 5 units left and 2 units down.

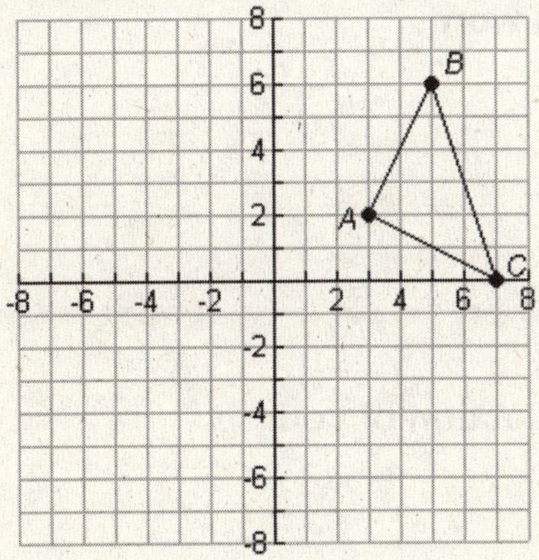

8. Rotate 90° clockwise.

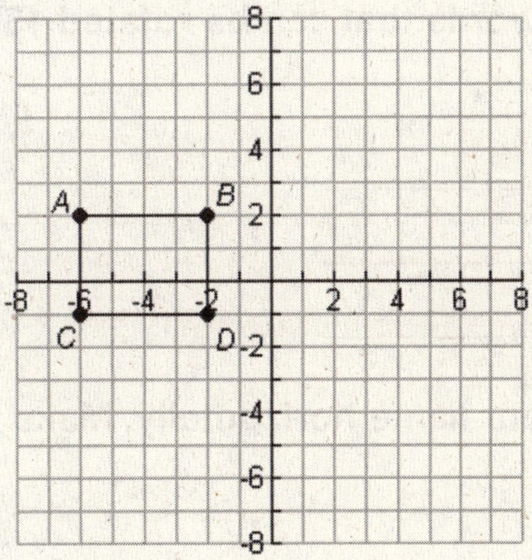

________________________ ________________________

Complete the figure. The dashed line is the line of symmetry.

9. 

10. 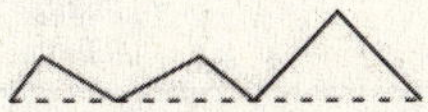

________________________ ________________________

Answers: **1.** Rectangle *ABCD* ≅ *EFGH* **2.** Trapezoid *CDEF* ≅ *HIJK* **3.** $r = 17$ **4.** $s = 100°$ **5.** $t = 87$ **6.** $u = 5°$ **7.** *A*' (−2, 0), *B*' (0, 4), *C*' (2, −2), **8.** *A*' (2, 6), *B*' (2, 2), *C*' (−1, 6), *D*' (−1, 2 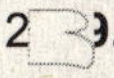 **9.** 

Name ______________________ Date ______________ Class ______________

CHAPTER 7

Family Fun

Transforming Words

What happens to some words or letters when they undergo a transformation? Which capital letters are symmetrical? Which words can be reflected and still stay the same? Which words change meanings when you rotate them?

Draw a line of symmetry on the capital letters that are symmetrical.

A B C D E F G H I J K L M

N O P R S T U V W X Y Z

Find 4 words which can be reflected and not lose their meaning.

____

____

____

____

Find 4 words that can be rotated 180° and still be a word.

____

____

____

____

Print your name horizontally. Make a horizontal reflection of your name.

____

Print your name vertically. Make a vertical reflection of your name.

____

Create a secret message in which the only way it can be understood is if it undergoes a transformation.

____

Answers: A B C D E H I K M O T U V W X Y sample: MOM WOW TAT MUM sample: MOM/WOW, ON/NO, MOW/MOW, NOON/NOON

Name ______________________ Date ______________ Class ______________

LESSON 7-6

Practice A

Congruence

Match each polygon in column A with a congruent polygon in column B.

Column A | **Column B**

1. 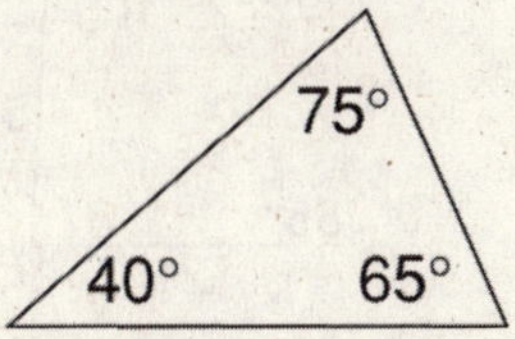

______

2. 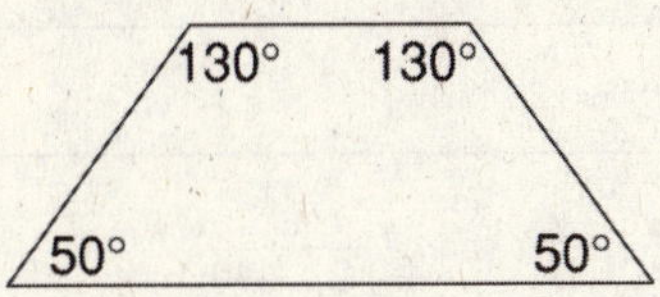

______

3. 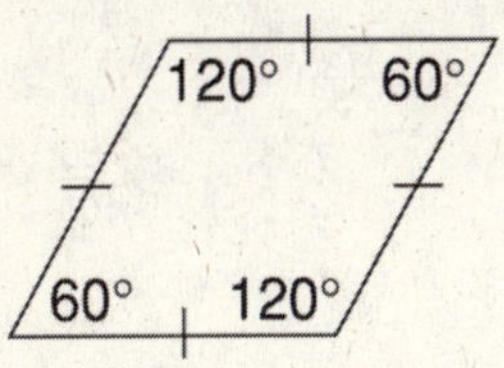

______

4. 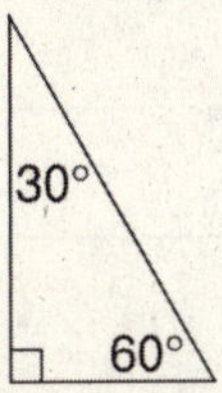

______

A. 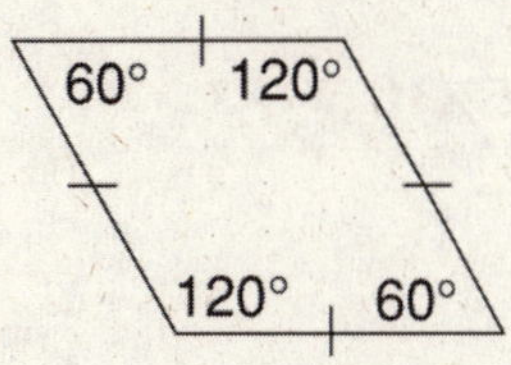

B. 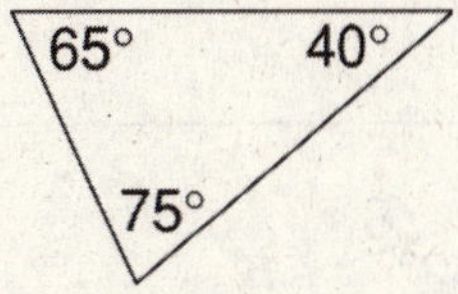

C. 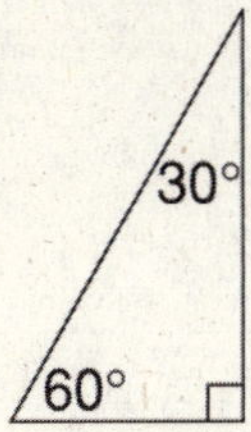

D. 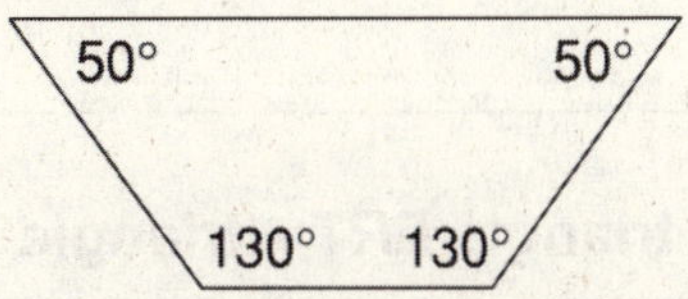

Write a congruence statement for each pair of congruent polygons.

5. 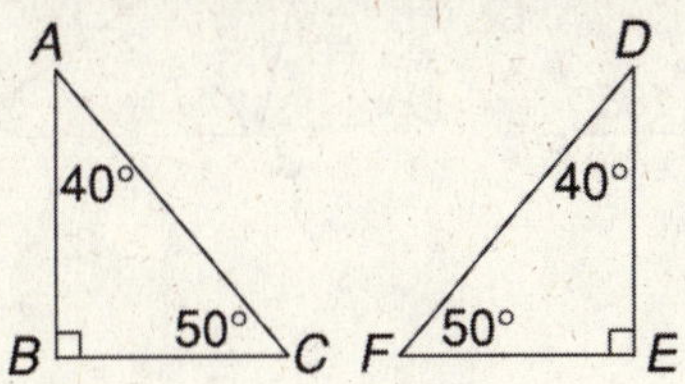

____

____

6. 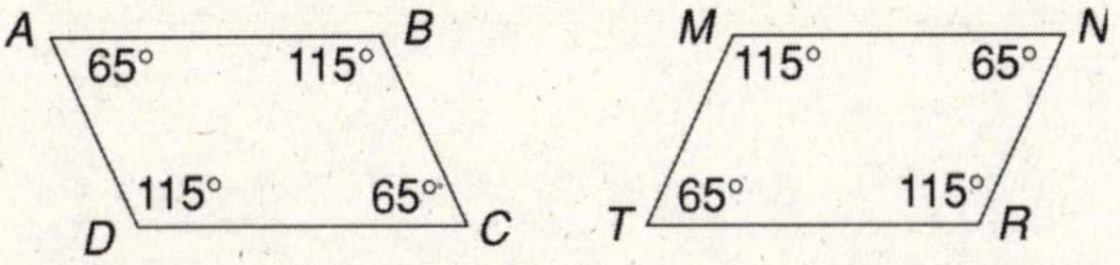

____

____

Triangle *ABC* is congruent to triangle *WXY*.

7. Find *d*. ______ 8. Find *t*. ______

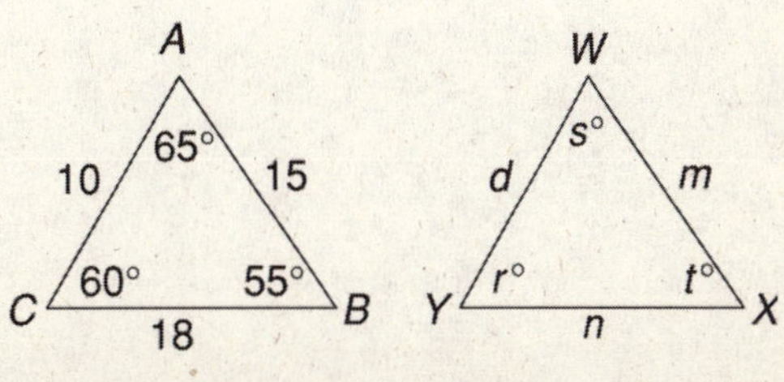

Name ______________________ Date ______________ Class ______________

LESSON 7-6

Practice B

Congruence

Write a congruence statement for each pair of congruent polygons.

1. 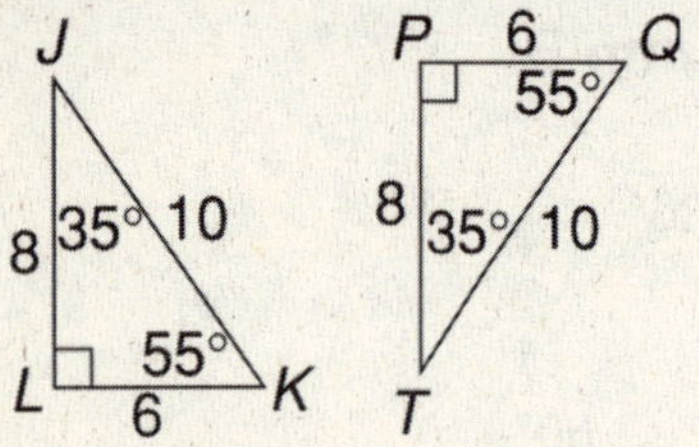

____

____

2. 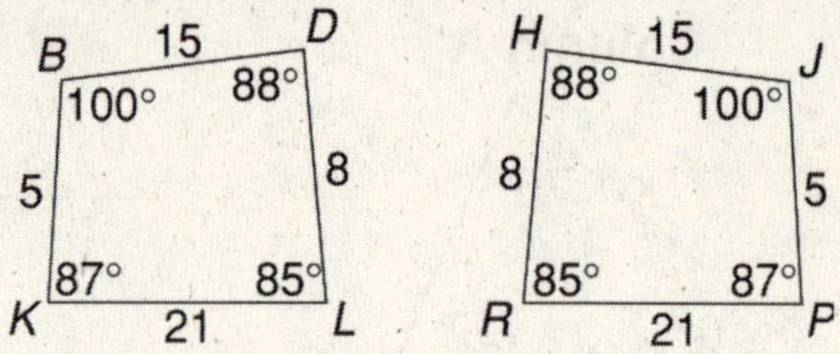

____

____

3. 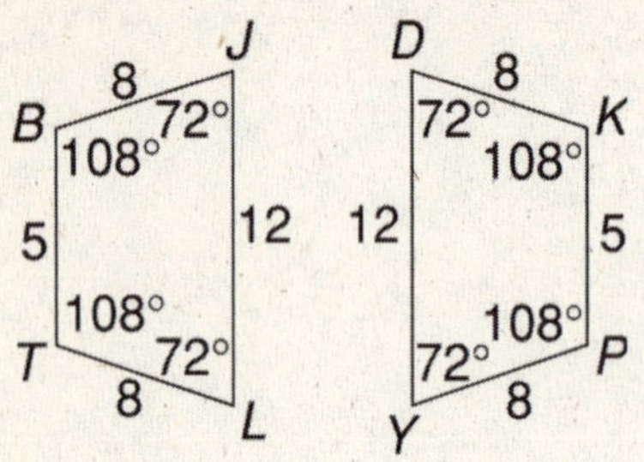

____

____

4. 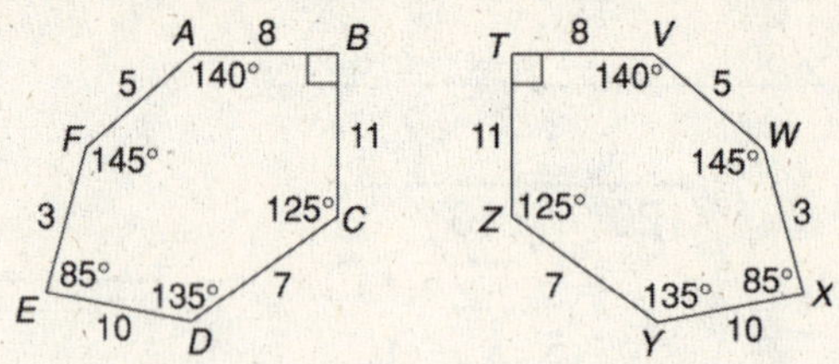

____

____

In the figure, triangle *PRT* ≅ triangle *FJH*.

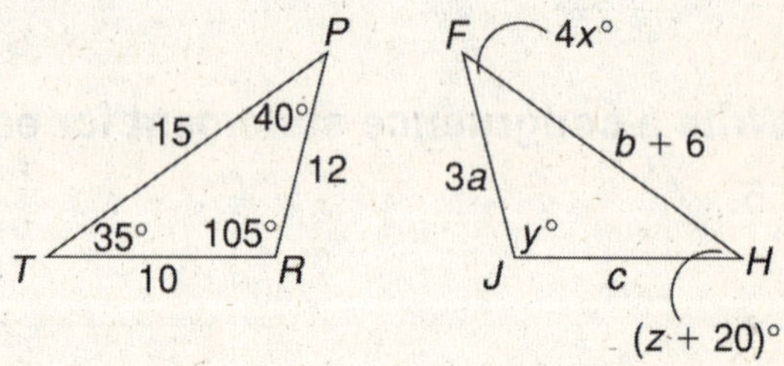

5. Find *a*. ____________

6. Find *b*. ____________

7. Find *c*. ____________

8. Find *x*. ____________

9. Find *y*. ____________

10. Find *z*. ____________

Name ______________________ Date ______________ Class ______________

LESSON 7-6

Practice C

Congruence

In the figure, quadrilateral *ABCD* ≅ quadrilateral *YZWX*.

1. Find *m*. ________

2. Find *h*. ________

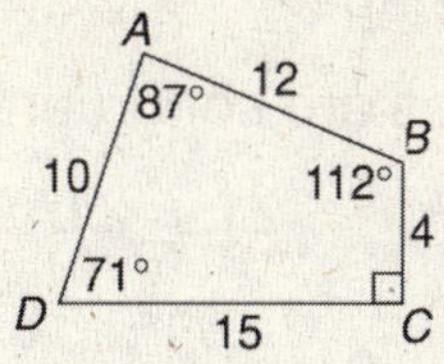

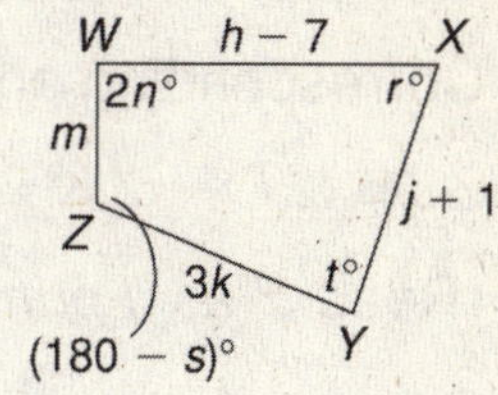

3. Find *j*. ________

4. Find *k*. ________

5. Find *n*. ________

6. Find *s*. ________

7. Find *t*. ________

8. Find *r*. ________

Find the value of the variables if octagon *ABCDEFGH* is congruent to octagon *VWXYZSTU*.

9. Find *a*. ________

10. Find *b*. ________

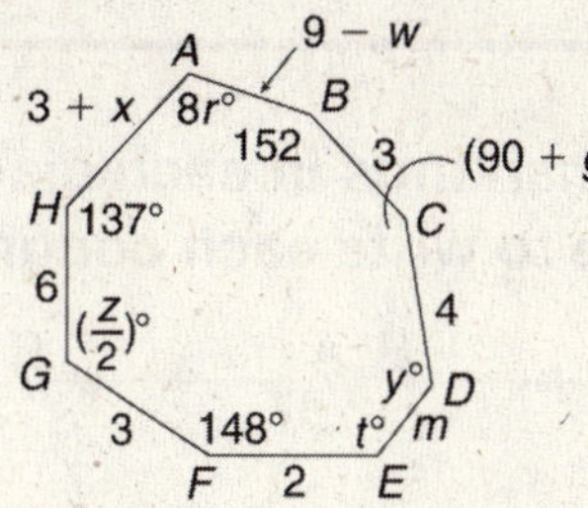

11. Find *c*. ________

12. Find *d*. ________

13. Find *g*. ________

14. Find *h*. ________

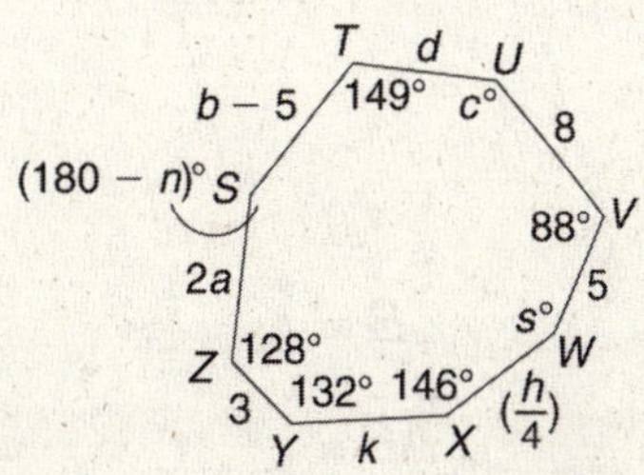

15. Find *k*. ________

16. Find *m*. ________

17. Find *n*. ________

18. Find *r*. ________

19. Find *s*. ________

20. Find *t*. ________

21. Find *w*. ________

22. Find *x*. ________

23. Find *y*. ________

24. Find *z*. ________

Name ______________________ Date ______________ Class ______________

LESSON 7-6

Review for Mastery

Congruence

Congruent polygons have the same size and shape.

Corresponding angles are congruent.

$\angle J \cong \angle J'$ $\angle K \cong \angle K'$ $\angle L \cong \angle L'$

(Read *J'* as *J prime*.)

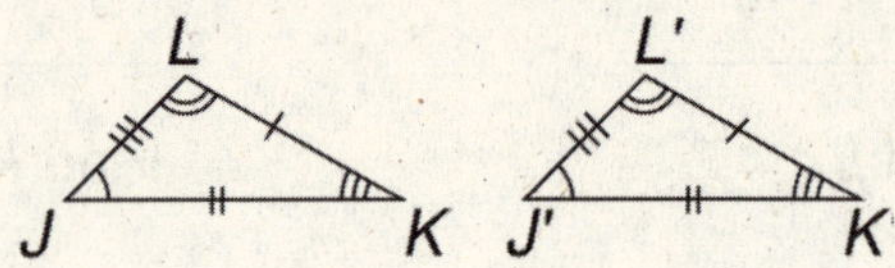

Corresponding sides are congruent.

$\overline{JK} \cong \overline{J'K'}$ $\overline{KL} \cong \overline{K'L'}$ $\overline{LJ} \cong \overline{L'J'}$

In a congruence statement, the vertices of the second polygon are written in order of correspondence with the first polygon. $\triangle JKL \cong \triangle J'K'L'$

Use the markings in each diagram.
Complete to write each congruence statement.

1.

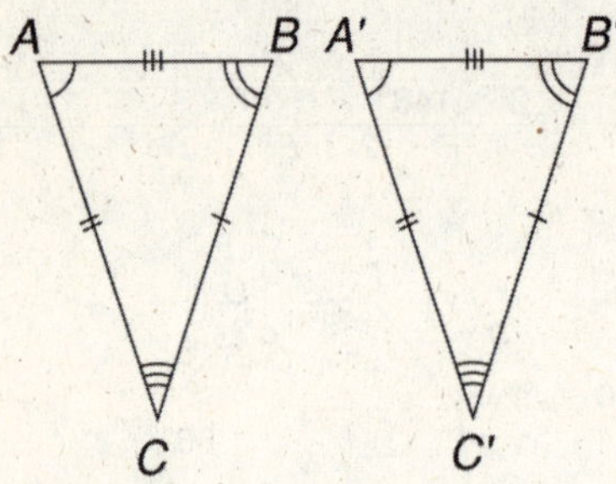

$\angle A \cong$ _____, $\angle B \cong$ _____,

$\angle C \cong$ _____, $\overline{AB} \cong$ _____,

$\overline{BC} \cong$ _____, $\overline{AC} \cong$ _____,

$\triangle ABC \cong$ __________

2.

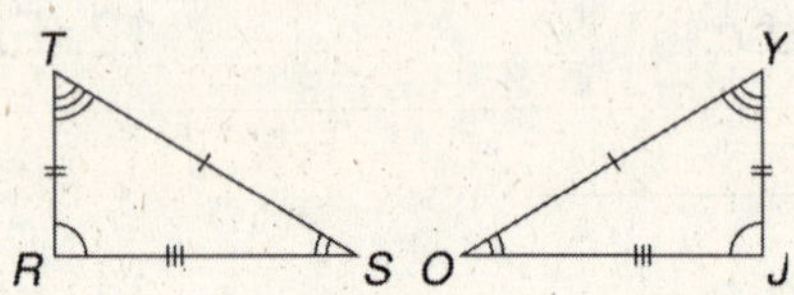

$\angle R \cong$ _____, $\angle S \cong$ _____,

$\angle T \cong$ _____, $\overline{RS} \cong$ _____,

$\overline{RT} \cong$ _____, $\overline{TS} \cong$ _____,

$\triangle RST \cong$ __________

3.

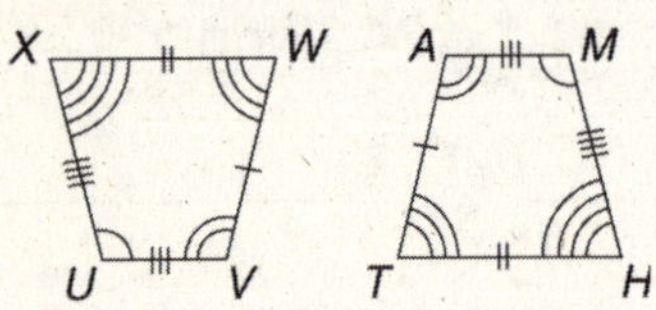

$\angle U \cong$ _____, $\angle V \cong$ _____, $\angle W \cong$ _____, $\angle X \cong$ _____,

$\overline{UV} \cong$ _____, $\overline{VW} \cong$ _____, $\overline{WX} \cong$ _____, $\overline{XU} \cong$ _____,

quad. *UVWX* $\cong$ __________

Name ______________________ Date ______________ Class ______________

LESSON 7-6

Review for Mastery

Congruence (continued)

Congruence relations can be used to find unknown values.

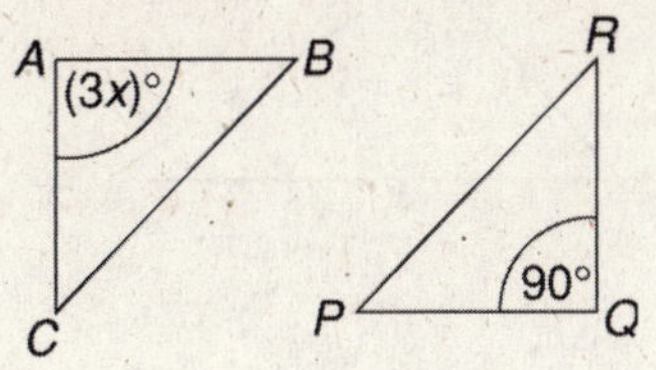

$\triangle ABC \cong \triangle QPR$

$$\angle A \cong \angle Q$$
$$3x = 90$$
$$\frac{3x}{3} = \frac{90}{3}$$
$$x = 30$$

Using the congruence relationship, complete to find each unknown value.

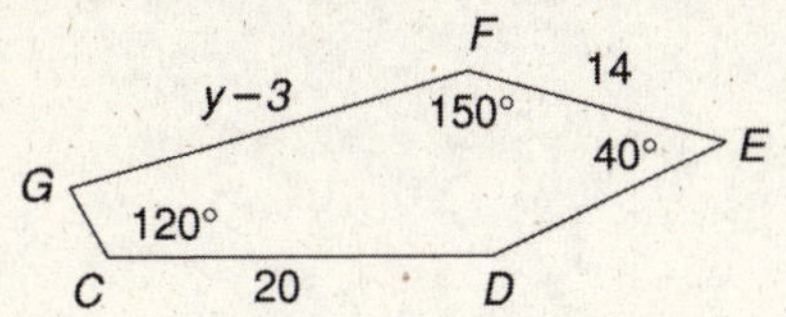

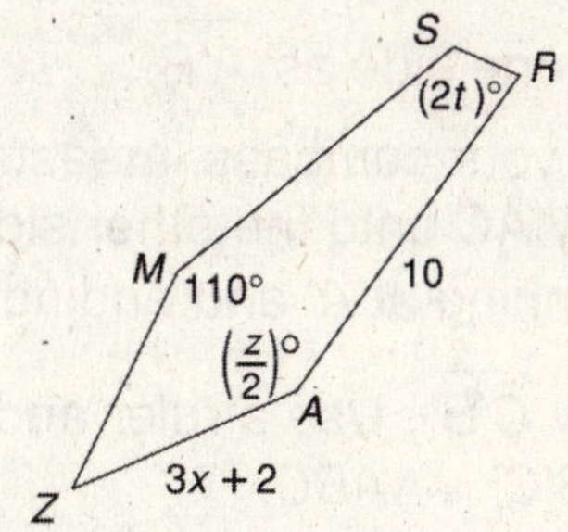

4. $\overline{GF} \cong \overline{RA}$

$y - 3 =$ ______

______ ______

$y =$ ______

5. $\angle C \cong$ ______

______ $= 2t$

______ $= \underline{2t}$

______ $= t$

6. $\angle A \cong$ ______

$\frac{z}{2} =$ ______

______ $\times \frac{z}{2} =$ ______ $\times$ ______

$z =$ ______

7. $\overline{AZ} \cong$ ______

$3x + 2 =$ ______

______ ______

$3x =$ ______

$\underline{3x} =$ ______

$x =$ ______

Name ____________________ Date ______________ Class ______________

LESSON 7-6

Challenge

Cloning

In the following exercises, you will construct a triangle congruent to ΔABC by copying three strategic parts.

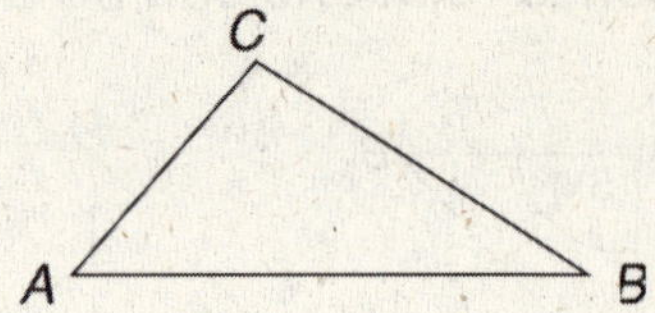

1. To construct a triangle congruent to ΔABC by using two pairs of sides and the included angle:

   a. With your compass, measure AB.
   Copy AB onto line ℓ.
   Call the copied length $A'B'$
   (read *A prime B prime*).

   b. With your compass, measure $\angle A$.
   Copy $\angle A$ at vertex A'
   with one side as $\overline{A'B'}$

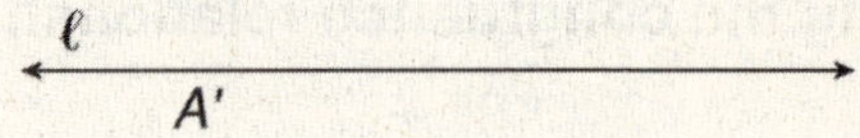

   c. With your compass, measure AC.
   Copy AC onto the other side of $\angle A'$,
   beginning at A' and ending at C'.

   d. Draw $\overline{C'B'}$. Use a ruler and protractor to verify that
   $\Delta A'B'C' \cong \Delta ABC$.

2. To construct a triangle congruent to ΔABC by using two pairs of angles and the included side:

   a. With your compass, measure AB.
   Copy AB onto line m.
   Call the copied length $A''B''$
   (*read A double prime B double prime*).

   b. With your compass, measure $\angle A$.
   Copy $\angle A$ at vertex A'', with one
   side as $\overline{A''B''}$.

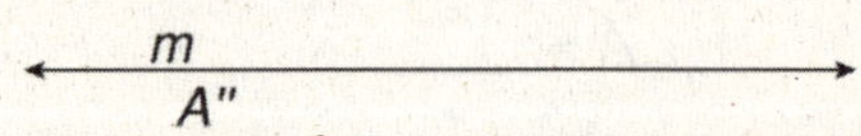

   c. With your compass, measure $\angle B$.
   Copy $\angle B$ at vertex B'', with one
   side as $\overline{A''B''}$.

   d. Use C'' to label the point where the sides of $\angle A''$ and $\angle B''$ intersect.
   Use a ruler and protractor to verify that $\Delta A''B''C'' \cong \Delta ABC$.

Name ______________________ Date ______________ Class ______________

LESSON 7-6

Problem Solving

Congruence

Use the American patchwork quilt block design called Carnival to answer the questions. Triangle *AIH* ≅ Triangle *AIB*, Triangle *ACJ* ≅ Triangle *AGJ*, Triangle *GFJ* ≅ Triangle *CDJ*.

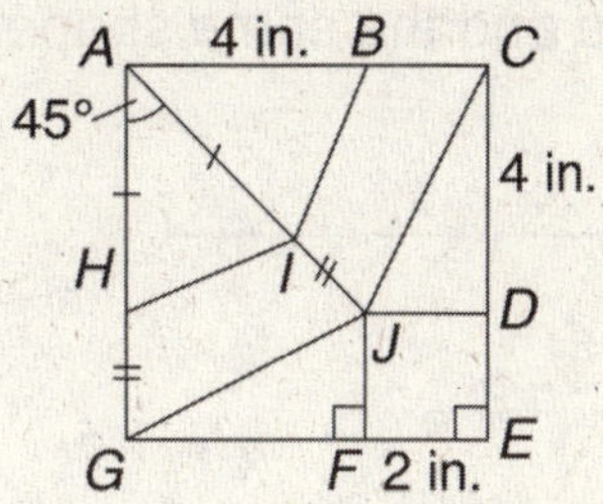

1. What is the measure of ∠*IAB*?

____

2. What is the measure of $\overline{AH}$?

____

3. What is the measure of $\overline{AG}$?

____

4. What is the measure of ∠*JDC*?

____

5. What is the measure of $\overline{FG}$?

____

The sketch is part of a bridge. Trapezoid *ABEF* ≅ Trapezoid *DEBC*. Choose the letter for the best answer.

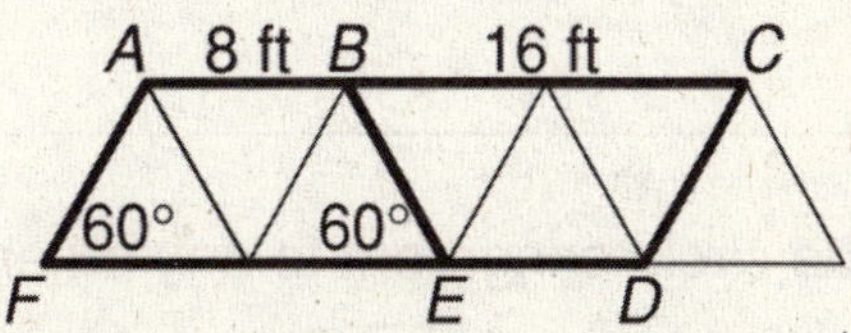

6. What is the measure of $\overline{DE}$?
   - A 4 feet
   - B 8 feet
   - C 16 feet
   - D Cannot be determined

7. What is the measure of $\overline{FE}$?
   - F 4 feet
   - H 8 feet
   - G 16 feet
   - J 24 feet

8. What is the measure of ∠*FAB*?
   - A 45°
   - C 60°
   - B 90°
   - D 120°

9. What is the measure of ∠*ABE*?
   - F 45°
   - H 60°
   - G 90°
   - J 120°

10. What is the measure of ∠*EBC*?
    - A 45°
    - C 60°
    - B 90°
    - D 120°

11. What is the measure of ∠*BED*?
    - F 45°
    - H 60°
    - G 90°
    - J 120°

12. What is the measure of ∠*BCD*?
    - A 45°
    - C 60°
    - B 90°
    - D 120°

Name ______________________ Date ______________ Class ______________

LESSON 7-6

Reading Strategies

Graphic Organizer

This picture helps you understand congruence.

Congruence

Definition
Two figures that have exactly the same size and the same shape.

Facts
- Figures have same size
- Figures have same shape

Examples

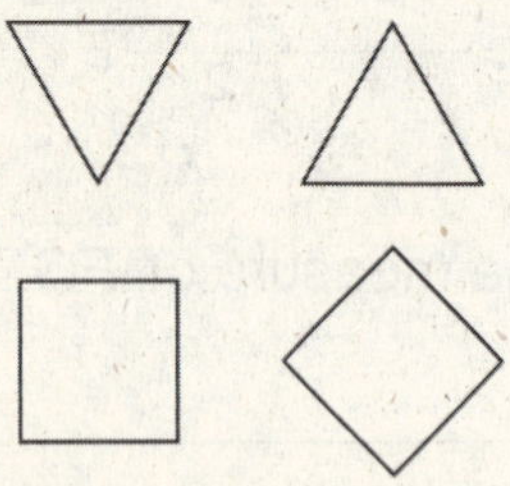

Non-examples

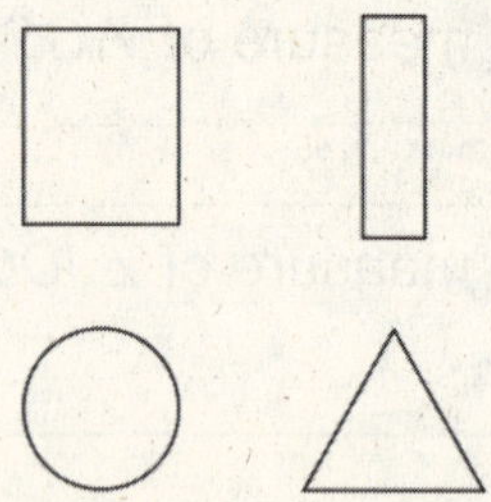

Use the picture to answer these questions.

1. How can you tell if two figures are congruent?

____

2. Is the circle congruent to any other shape? ______________

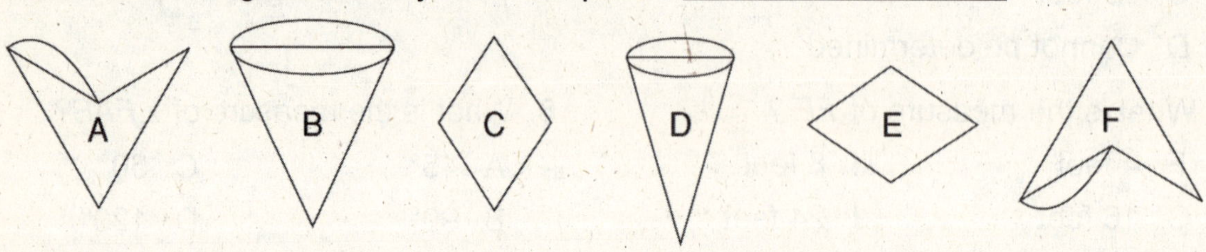

Use the chart and the figures above to answer the following questions.

3. Is figure A congruent to figure B? ______________

4. Is figure B congruent to figure D? ______________

5. Is figure C congruent to figure E? ______________

6. Is figure A congruent to figure F? ______________

Name ______________________ Date ______________ Class ______________

LESSON 7-6

Puzzles, Twisters & Teasers

Equal Time!

In each problem there are congruent figures with three answers. Decide which answer is correct. Write the letter of the correct answer on the blank line which corresponds to the problem.

1. **K** $\Delta KLM \cong \Delta RQS$
   R $\angle K \cong \angle Q$
   M $\overline{KM} \cong \Delta \overline{QS}$

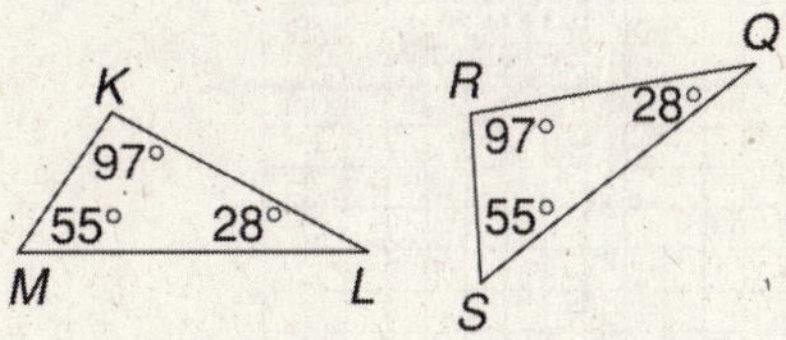

2. **T** $\overline{ED} \cong \overline{JF}$
   S $\angle B \cong \angle H$
   F pentagon *ABCDE* ≅ pentagon *JIHGF*

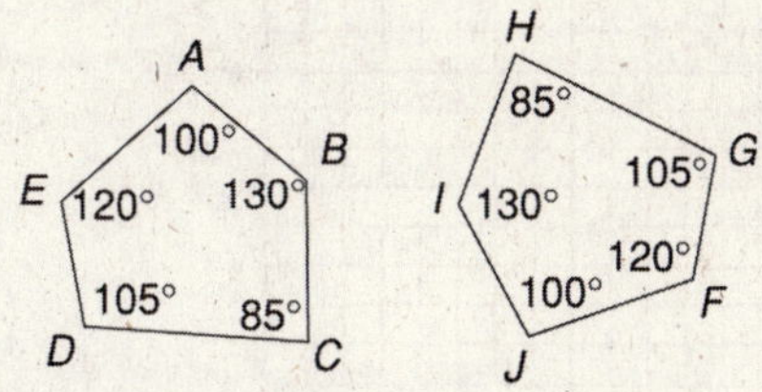

3. **E** $\overline{AB} \cong \overline{FE}$
   G $\Delta ACB \cong \Delta DEF$
   W $\Delta CAB \cong \Delta EFD$

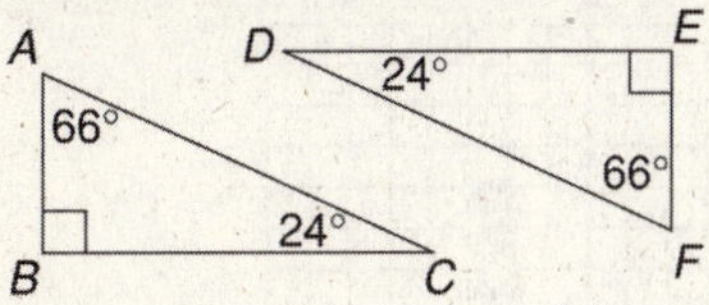

4. **K** quadrilateral *LMNO* ≅ quadrilateral *QRST*
   L $\angle N \cong \angle T$
   S $\angle L \cong \angle T$

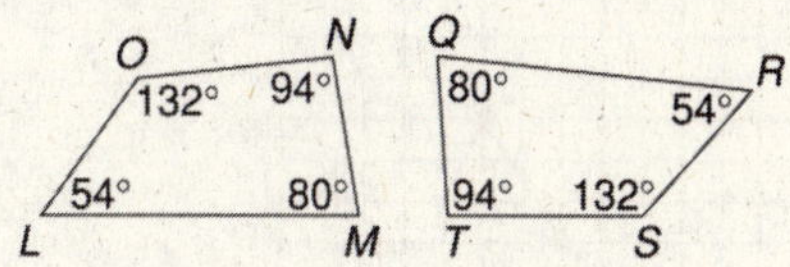

5. **C** $\Delta CAB \cong \Delta JKL$
   D $\angle A \cong \angle L$
   N $ACB \cong KLJ$

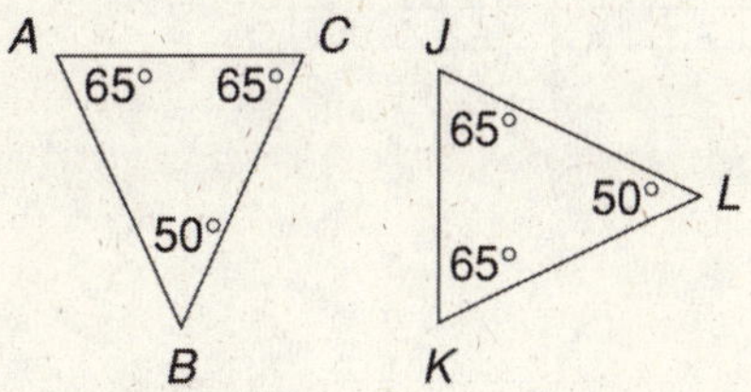

6. **I** quadrilateral *PQSR* ≅ quadrilateral *TUVW*
   A $\angle P \cong \angle U$
   O $\overline{RP} \cong \overline{TW}$

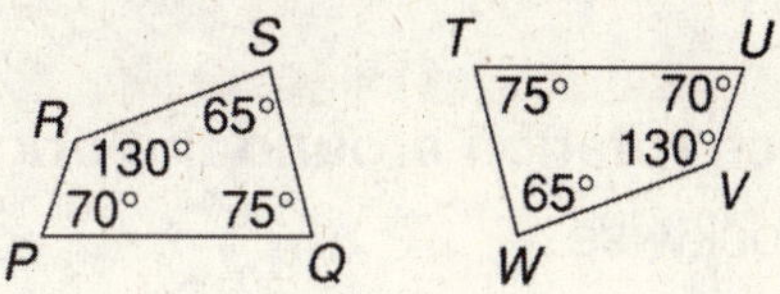

7. **E** $\angle D \cong \angle L$
   A $\overline{AD} \cong \overline{ML}$
   O $\overline{AB} \cong \overline{NM}$

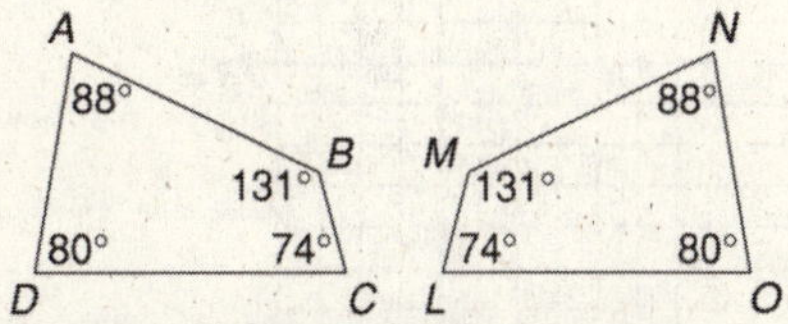

What kind of face has hands but no eyes, nose, or mouth?

A ____ ____ ____ ____ ____   ____ ____ ____ ____

5 4 7 5 1   2 6 5 3

Name ____________________ Date ____________ Class ____________

LESSON 7-7

Practice A

Transformations

Graph each translation.

1. 3 units right and 4 units down

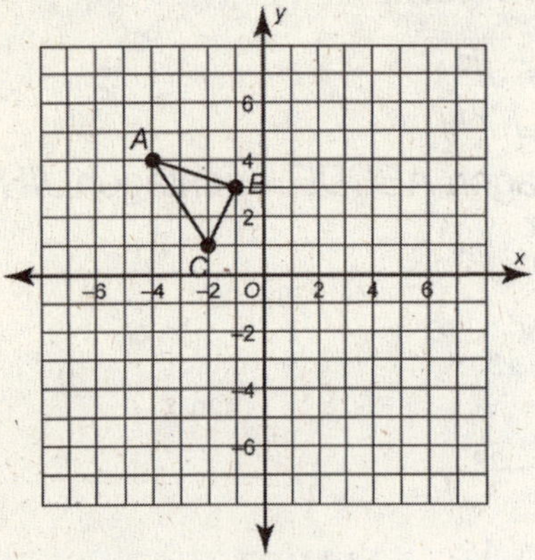

2. 5 units left and 2 units up

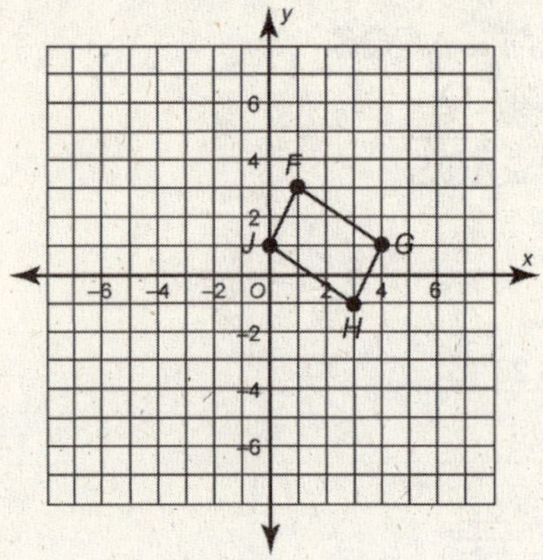

Graph each reflection.

3. across the *x*-axis

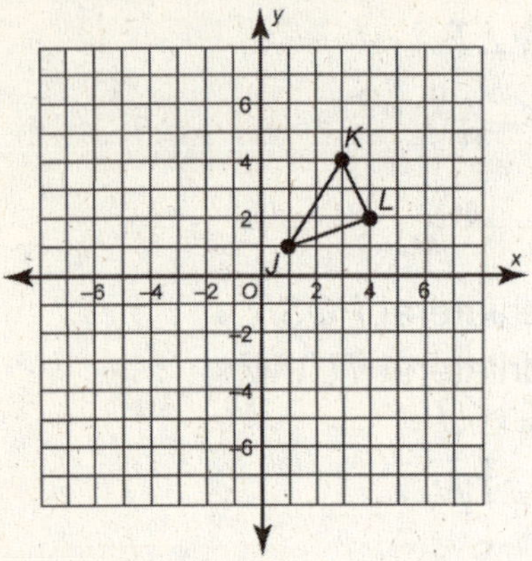

4. across the *y*-axis

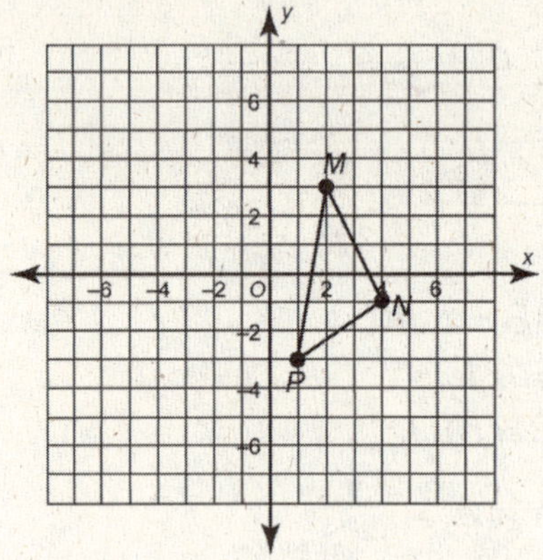

Graph each rotation around the origin.

5. 90° clockwise

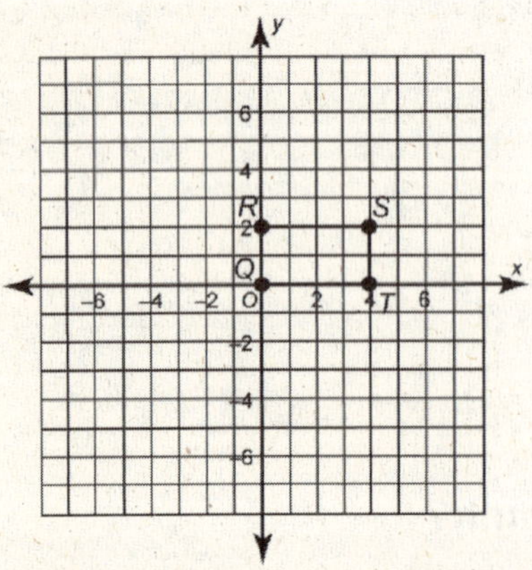

6. 180°

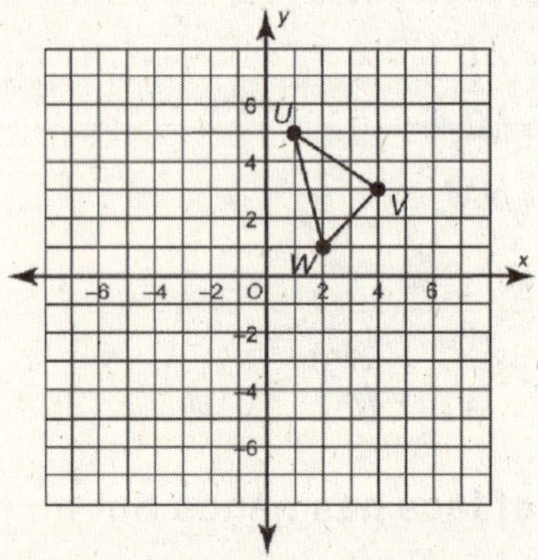

7. A triangle has vertices *X*(–7, 4), *Y*(–3, 1), and *Z*(0, 5). After a transformation, the coordinates of the image are *X*' (–1, 1), *Y*' (3, –2), and *Z*' (6, 2). Describe the transformation. ____________________

Name ______________________ Date ______________ Class ______________

LESSON 7-7

Practice B

Transformations

Graph each translation.

1. 3 units left and 9 units down

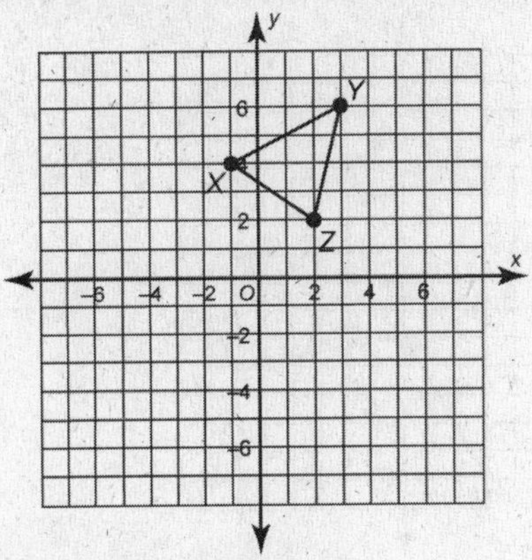

2. 3 units right and 6 units up

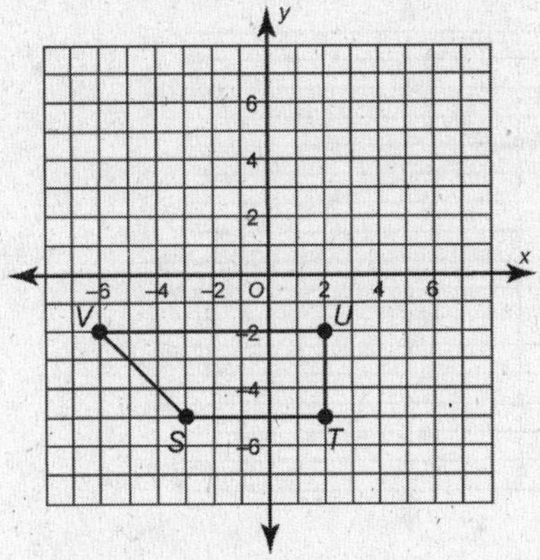

Graph each reflection.

3. across the *x*-axis

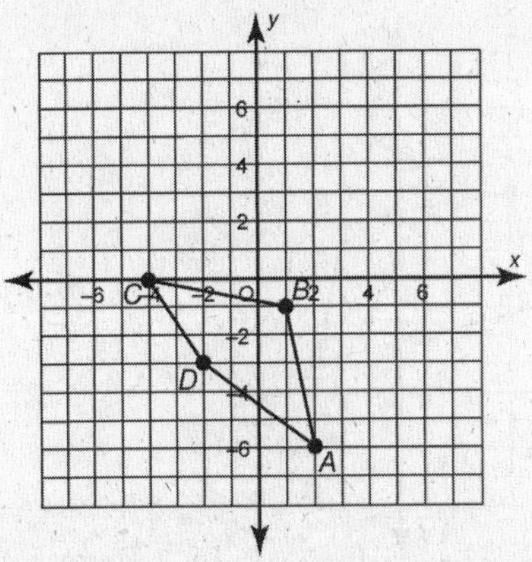

4. across the *y*-axis

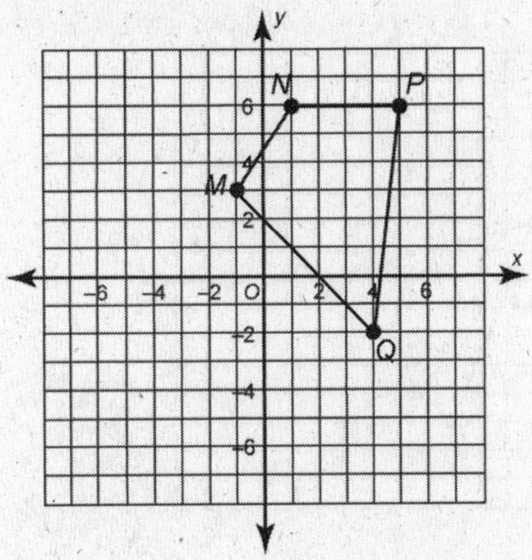

Graph each rotation around the origin.

5. 90° clockwise

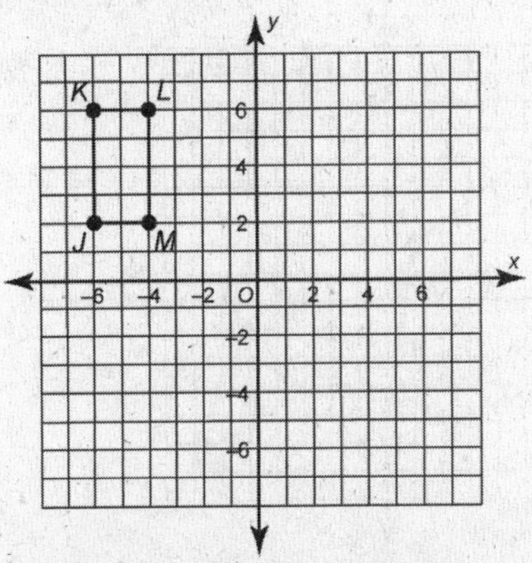

6. 180°

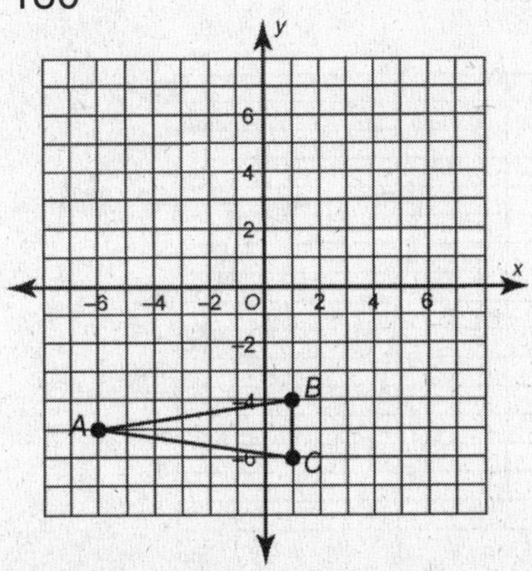

7. A parallelogram has vertices *A*(–1, 3), *B*(4, 3), *C*(6, –1), and *D*(1, –1). After a transformation, the coordinates of the image are *A'* (1, 3), *B'* (–4, 3), *C'* (–6, –1), and *D'* (–1, –1). Describe the transformation. ______________________

Name ______________________ Date ______________ Class ______________

LESSON 7-7

Practice C

Transformations

Graph each translation.

1. 1 unit left and 6 units down

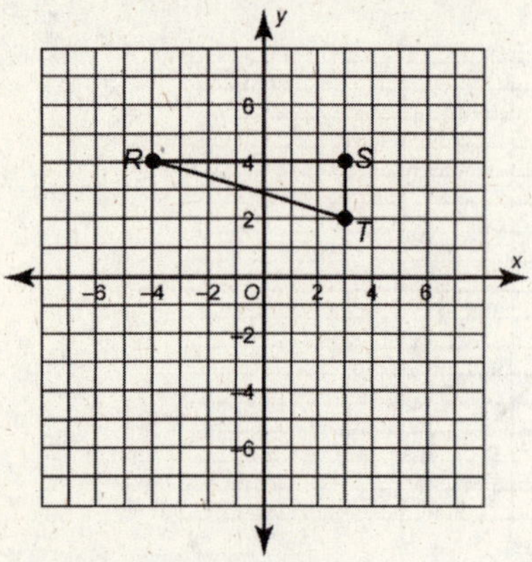

2. 3 units left and 7 units up

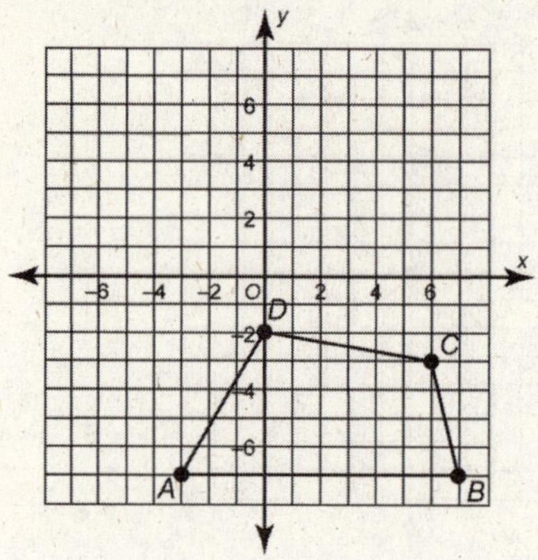

Graph each reflection.

3. across the *x*-axis

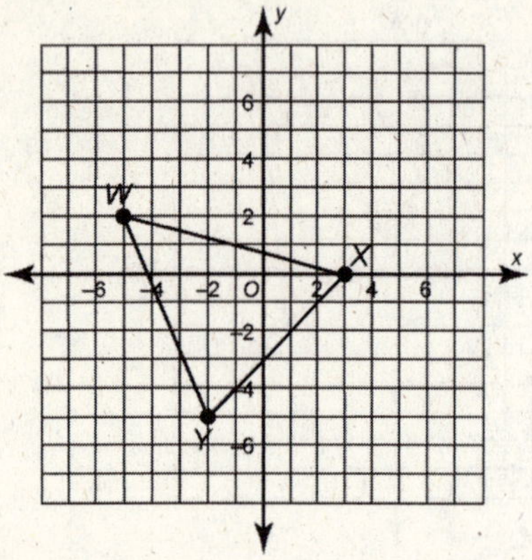

4. across the *y*-axis

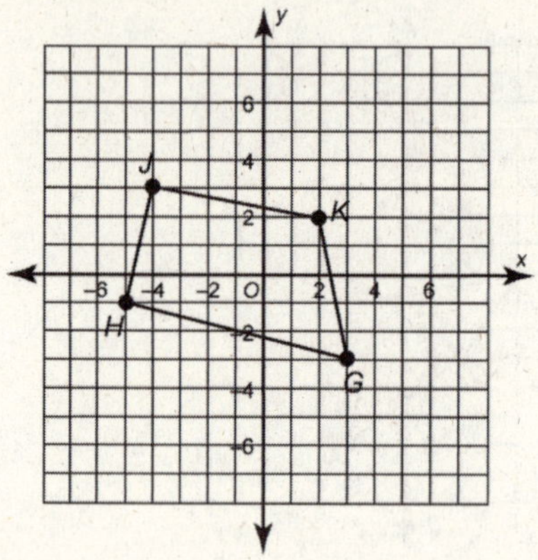

Graph each rotation around the origin.

5. 90° clockwise

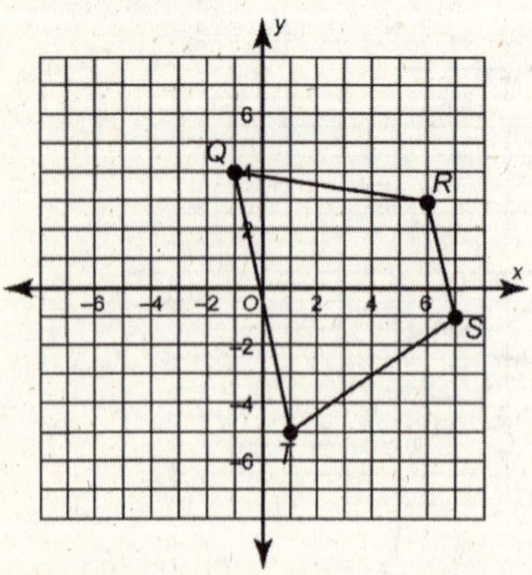

6. 180°

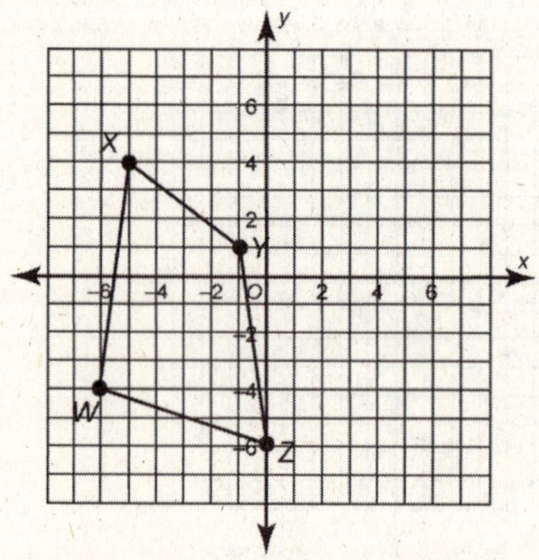

7. A trapezoid has vertices *Q*(−2, 5), *R*(3, 5), *S*(6, 1), and *T*(−2, 1). After a transformation, the coordinates of the image are *Q*' (−5, −2), *R*' (−5, 3), *S*' (−1, −2), and *T*' (−1, 6). Describe the transformation. ______________________

Name ______________________ Date ______________ Class ______________

LESSON 7-7

Review for Mastery

Transformations

Transforming a preimage means moving the figure to a new location on a coordinate grid according to a rule. The resulting figure is the image. The image of point *A* is noted as *A*' .

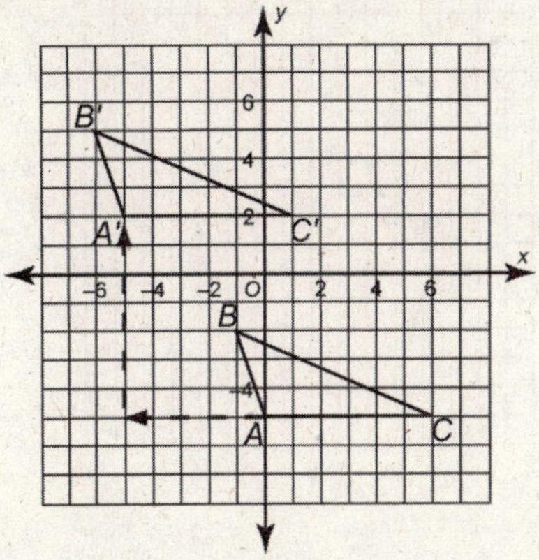

A translation is a slide along a horizontal line and then a vertical line. For each vertex of the preimage, count to locate the coordinates of the image of that vertex. Triangle *ABC* is translated 5 units left and 7 units up.

Graph each translation.

1. 5 units right and 1 unit down

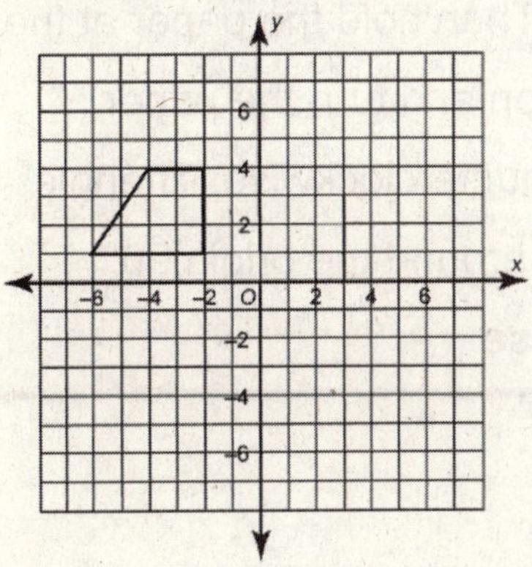

2. 3 units left and 4 units up

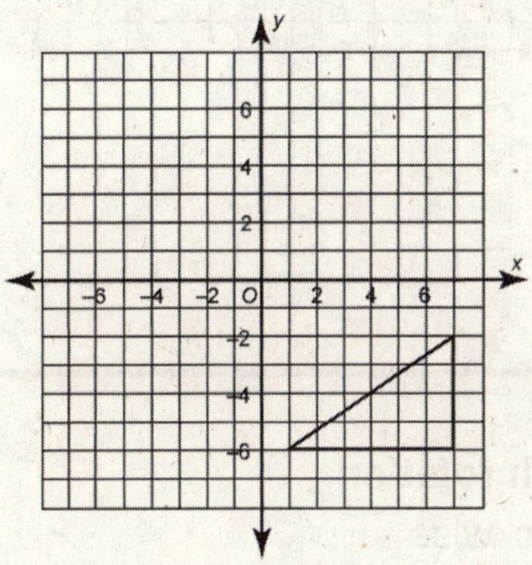

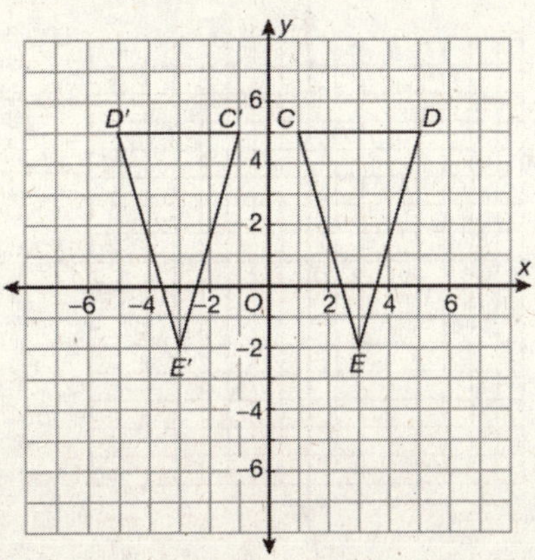

A reflection is a flip, or mirror image across a line. Each vertex of the preimage, is located the same number of units from the line of reflection, but in the opposite direction. Triangle *CDE* is reflection across the *y*-axis.

Name ______________________ Date ______________ Class ______________

LESSON 7-7

Review for Mastery

Transformations (continued)

Graph each reflection.

3. across the *y*-axis

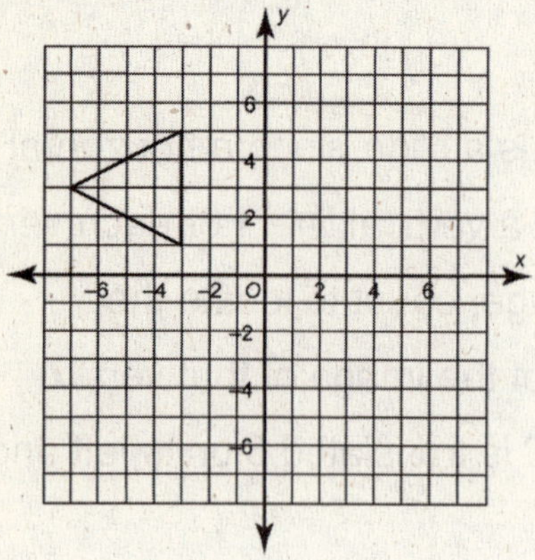

4. across the *x*-axis

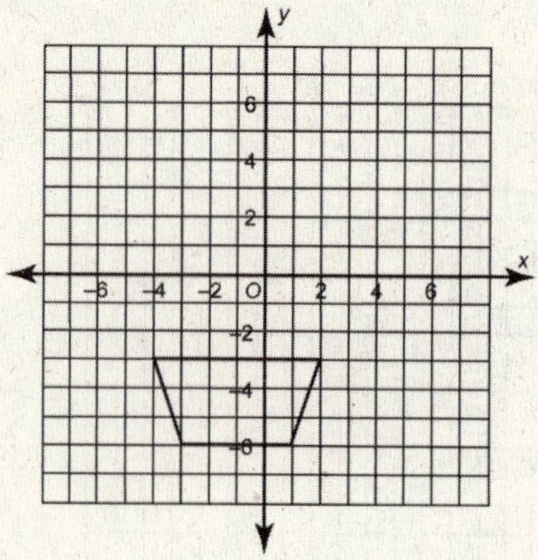

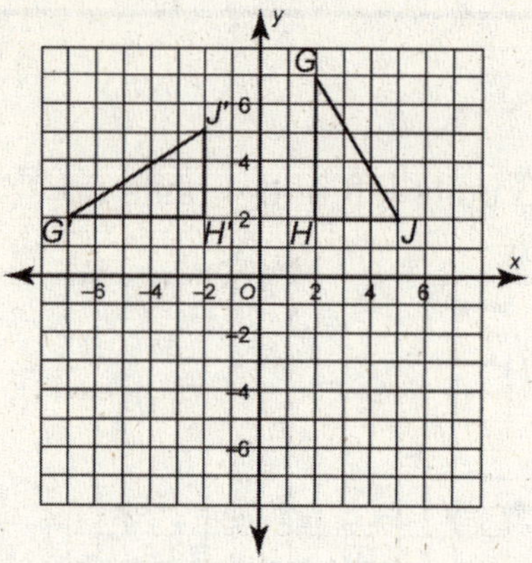

A rotation is a turn around a point. Lay tracing paper over the grid and draw the preimage. Then hold the paper at the center of rotation and turn the paper clockwise or counterclockwise. Triangle *GHJ* is rotated around the origin 90° counterclockwise.

Graph each rotation.

5. 90° clockwise

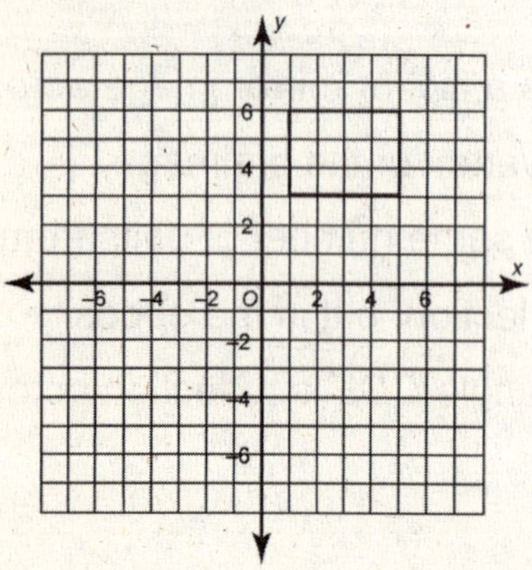

6. 180°

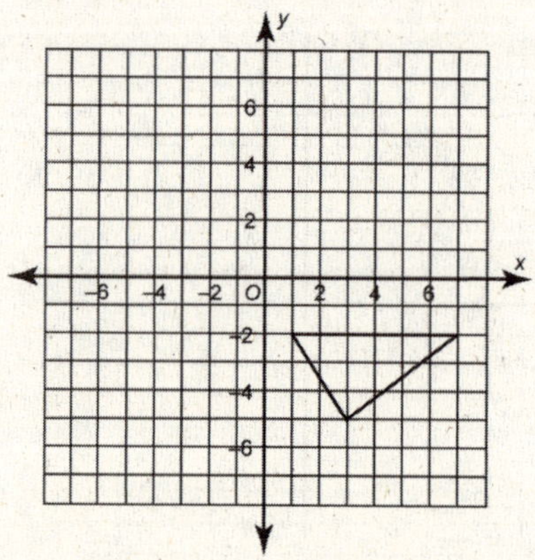

Name ______________________ Date ______________ Class ______________

LESSON 7-7

Challenge

Tessellations

A **tessellation** is a pattern of a figure that repeats and covers a plane without overlapping and without gaps.

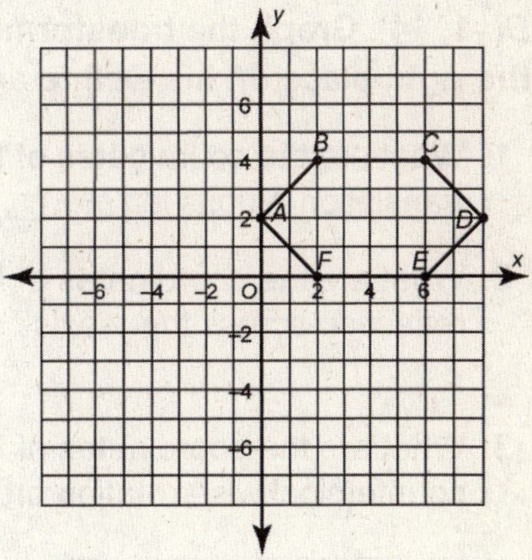

1. Translate figure *ABCDEF* 6 units left and 2 units down. What are the coordinates of *A′B′C′D′E′F′* ?

____

2. Translate figure *A′B′C′D′E′F′* 4 units down. What are the coordinates of *A″B″C″D″E″F″* ?

____

3. Translate figure *A″B″C″D″E″F″* 6 units right and 2 units up. What are the coordinates of *A‴B‴C‴D‴E‴F‴* ?

____

4. Does a translation of figure *ABCDEF* result in a tessellation? Explain.

____

____

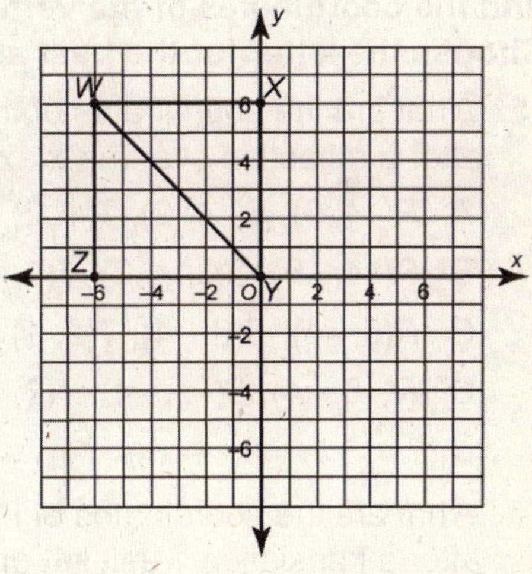

5. Rotate figure *WXYZ* 90° clockwise. What are the coordinates of *W′X′Y′Z′* ?

____

6. Rotate figure *WXYZ* 180°. What are the coordinates of *W″X″Y″Z″* ?

____

7. Rotate figure *WXYZ* 90° counterclockwise. What are the coordinates of *W‴X‴Y‴Z‴* ?

____

8. Does a rotation of figure *WXYZ* result in a tessellation? Explain.

____

____

9. Can you recreate the image formed above using a different transformation? Explain.

____

____

Name ______________________ Date ______________ Class ______________

LESSON 7-7

Problem Solving

Transformations

Drew is going to hang a flag in his bedroom that is in the shape of a trapezoid. The grid below represents Drew's wall. The flag, a trapezoid *ABCD*, has vertices *A*(–1, –1), *B*(2, –1), *C*(4, –4), and *D*(–1, –4). Graph the transformations that Drew performs to find the right place on his wall to answer the questions.

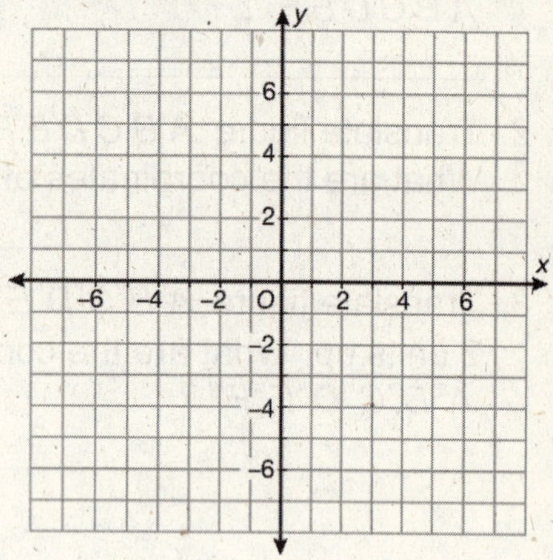

1. What are the coordinates of Point *A* after a translation 4 units left? ______________

2. What are the coordinates of Point *B* after a reflection across the *y*-axis?

   ____

3. What are the coordinates of Point *C* after a 90° counterclockwise rotation around the origin?

   ____

4. What are the coordinates of Point *D* after a 180° clockwise rotation around the origin?

   ____

A blueprint of a sailboat has a sail in the shape of a right triangle, *RST*. On the blueprint the vertices are *R*(–4, 6), *S*(–4, 2) and *T*(1, 2). Use the rules to find the coordinates of the vertices of the sailboat as it changes position. Choose the letter for the best answer.

5. What are the coordinates of the sail after a reflection over the *x*-axis?

   A *R*(–4, 6), *S*(–4, 2), *T*(1, 2)

   B *R*(–4, –6), *S*(–4, –2), *T*(1, –2)

   C *R*(6, –4), *S*(2, –4), *T*(2, 1)

   D *R*(–6, –4), *S*(–2, –4), *T*(2, 1)

6. What are the coordinates of the sail after a translation left 3 units and a reflection over the *y*-axis?

   F *R*(7, 6), *S*(7, 2), *T*(2, 2)

   G *R*(7, –6), *S*(7, –2), *T*(2, –2)

   H *R*(–7, –6), *S*(–7, –2), *T*(–2, –2)

   J *R*(–6, –7), *S*(–2, –7), *T*(–2, –2)

7. What are the coordinates of Point *R* after a translation 1 unit left and a 90° clockwise rotation about (0, 0)?

   A *R*(5, 6) C *R*(–5, 6)

   B *R*(–6, –5) D *R*(6, 5)

8. What are the coordinates of Point *S* after a translation 4 units to the right and 5 units down?

   F *S*(–3, 0) H *S*(0, 3)

   G *S*(0, –3) J *S*(3, 0)

Name ____________________ Date ____________ Class ____________

LESSON 7-7

Reading Strategies

Understanding Vocabulary

A **transformation** is a change in a figure's position or size. A figure that is transformed always keeps the same shape.

There are three transformations that move a figure to a new position. The size stays the same. A **translation** is a **slide**, a **rotation** is a **turn**, and a **reflection** is a **flip**.

Translation → **Slide** a figure to a new position.	**Reflection** → **Flip** a figure over a line to a new position.	**Rotation** → **Turn** a figure around one point to a new position.
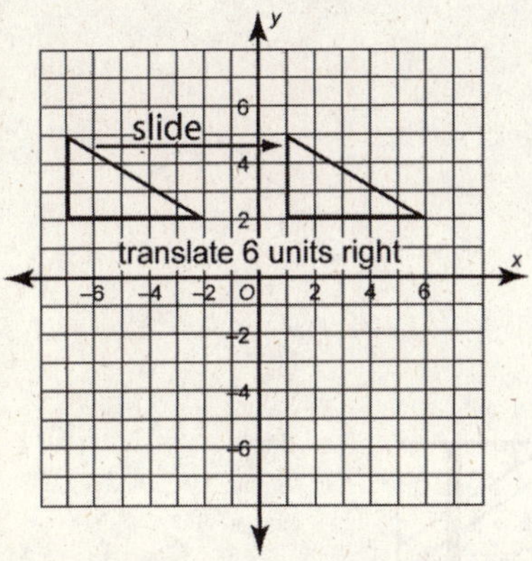	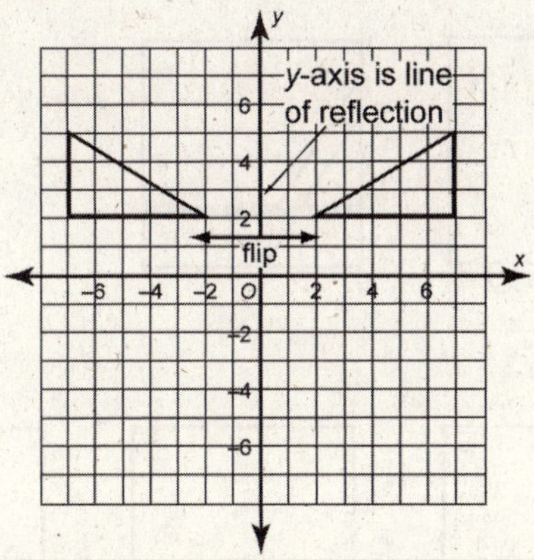	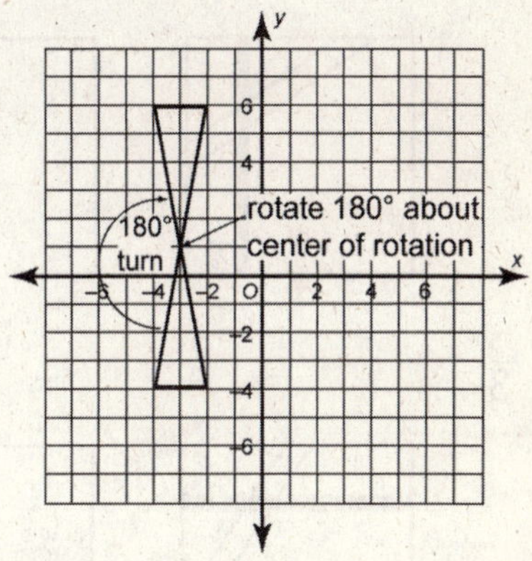

Write the name of each kind of transformation: *translation*, *rotation*, or *reflection*.

1.

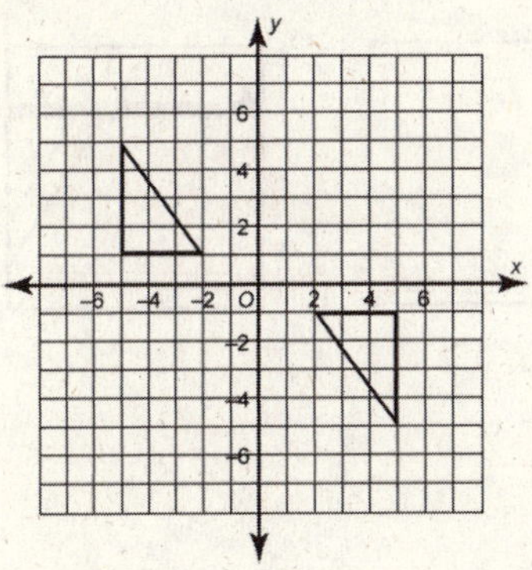

____

2.

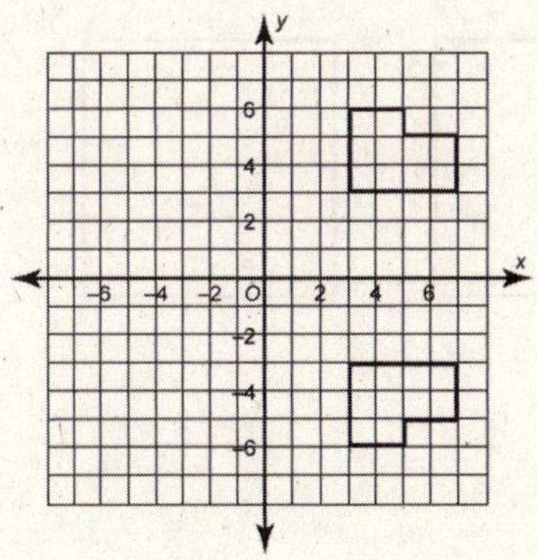

____

3.

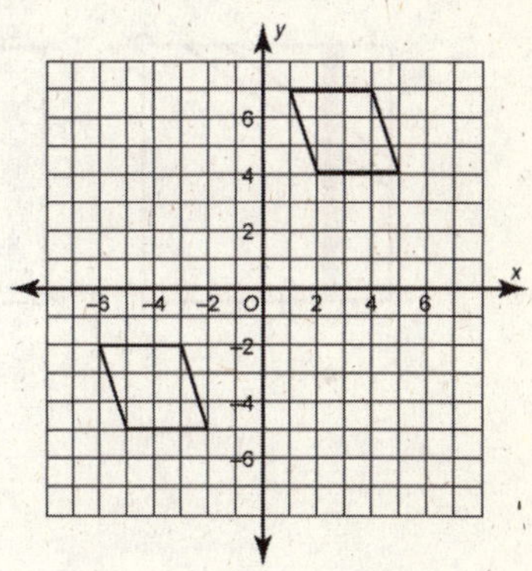

____

Name ______________________ Date ______________ Class ______________

LESSON 7-7

Puzzles, Twisters & Teasers

See Those Shapes Move

For each row, transform the picture on the left to look like the picture on the right. Each blank box represents a step. Sketch and name the transformation in each step.

1.

2.

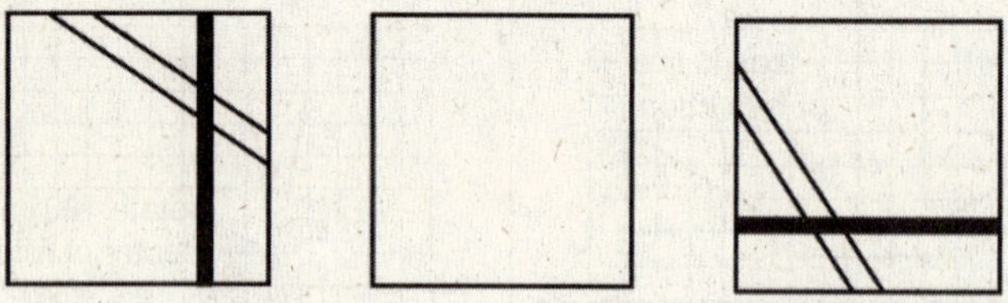

3.

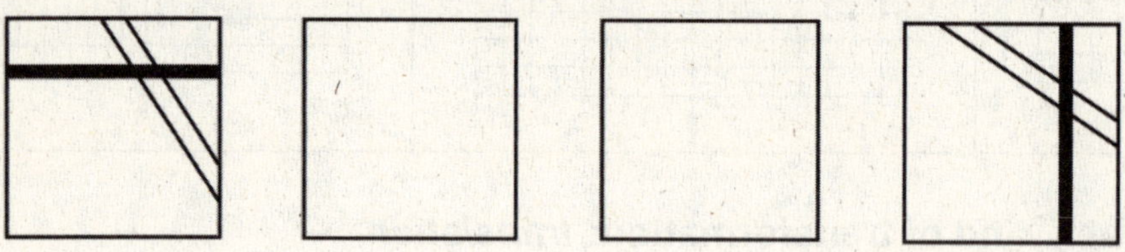

4.

Name ______________________ Date ______________ Class ______________

LESSON 7-8

Practice A
Symmetry

Complete the figure. The dashed line is the line of symmetry.

1. 

2. 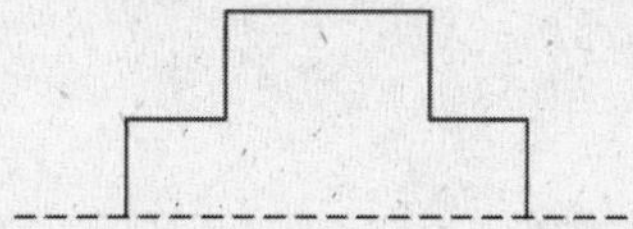

3. 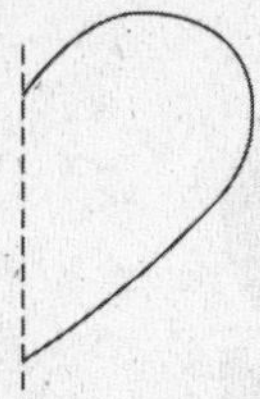

4. 

5. 

6. 

Complete the figure. The point is the center of rotation.

7. 4-fold

8. 2-fold

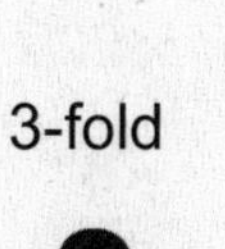

9. 5-fold

10. 3-fold

Name ______________________ Date ______________ Class ______________

LESSON 7-8

Practice B

Symmetry

Complete each figure. The dashed line is the line of symmetry.

1.

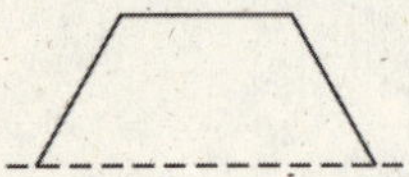

2.

3.

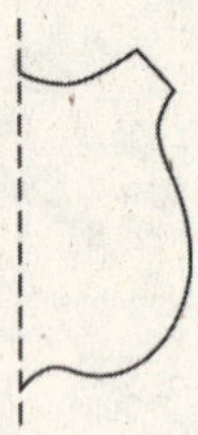

4.

5.

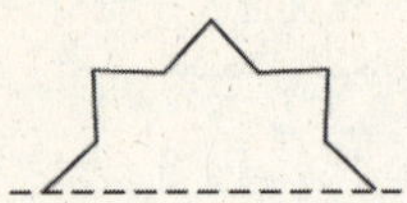

6.

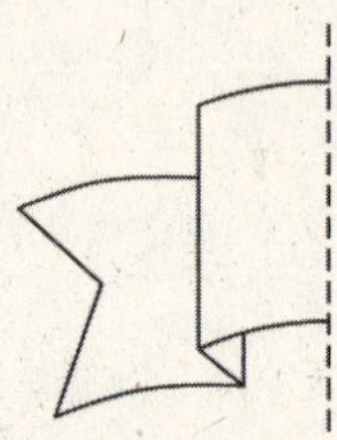

Complete each figure. The point is the center of rotation.

7. 5-fold

8. 4-fold

9. 2-fold

10. 2-fold

Name ______________________________ Date ________________ Class ________________

LESSON 7-8

Practice C

Symmetry

Complete each figure. The dashed line is the line of symmetry.

1.

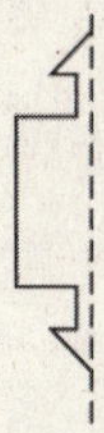

2.

3.

4.

Complete each figure. The point is the center of rotation.

5. 3-fold

6. 5-fold

Draw an example of a figure with each type of symmetry.

7. line symmetry and rotational symmetry

8. line symmetry but not rotational symmetry

Name ____________________ Date ______________ Class ______________

LESSON 7-8

Review for Mastery

Symmetry

Vertical Line Symmetry	Horizontal Line Symmetry	Diagonal Line Symmetry
	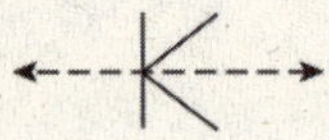	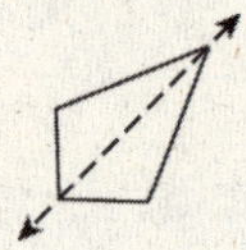

Draw all the lines of symmetry in each figure.

1. 

2. 

3. 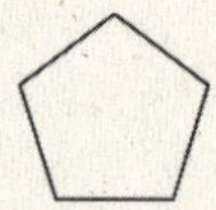

Rotational symmetry is when you can turn an object so that it looks exactly the same. The number of positions in which it looks exactly the same gives you its order of symmetry.

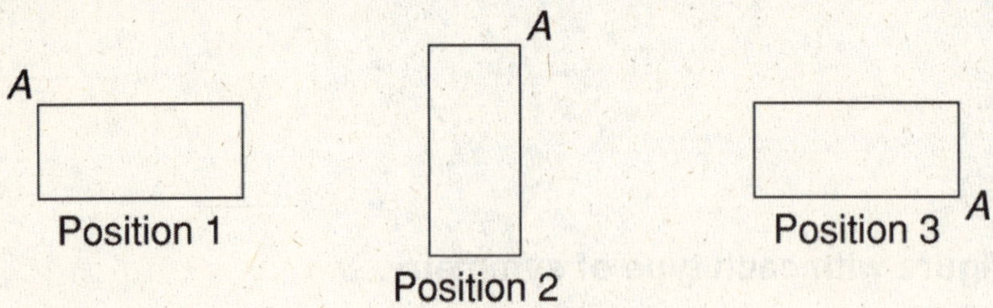

Follow vertex *A* to see how the figure is turned.
Just twice, the figure looks exactly the same, in Positions 1 and 3.
So, this figure has 2-fold rotational symmetry.

Tell the order of rotational symmetry.

4. 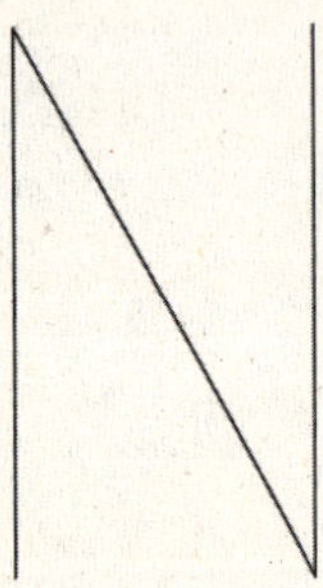

____

5. 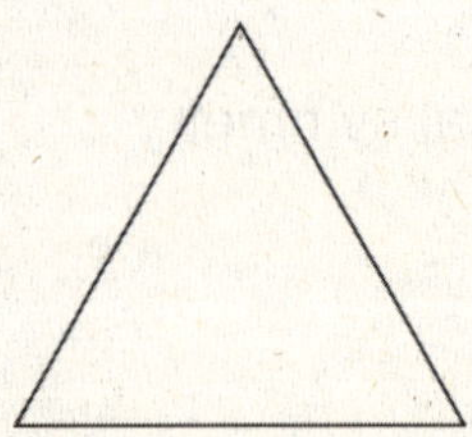

____

6. 

____

Name ______________________ Date ______________ Class ______________

LESSON 7-8

Challenge

Inside, Outside

Materials: sheet of paper (8.5 in. by 11 in.), scissors, tape

1. a. Cut a strip of paper 2 in. by 11 in. Tape the ends together to form a loop.

   b. Use a pencil to create a center line around the outside of the loop. Cut the loop along its center line. How many loops? __________

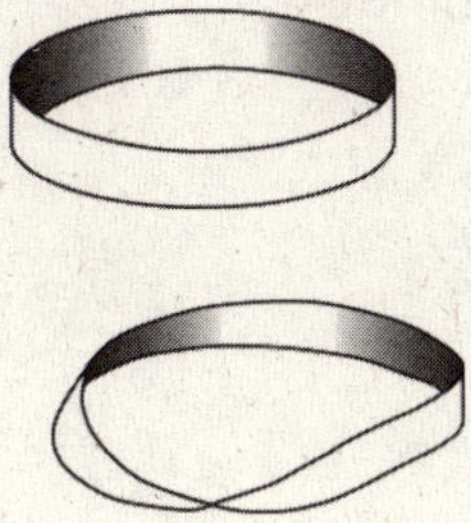

2. a. Cut a strip of paper 2 in. by 11 in. Before taping the ends together give the strip a half-twist by turning one end over 180°

   b. Use a pencil to create a center line around the loop. Continue drawing until you reach your starting point. Describe the result. How many surfaces (sides) does this strip have?

   ____

   ____

   c. Cut the loop along its center line. How many loops do you have?

   ____

This unusual surface is called a **Mobius Strip,** named after A. F. Mobius, a 19th-century German mathematician and astronomer, who pioneered *topology* (how geometric figures act when distorted by such things as twists and stretches).

3. Prepare another Mobius Strip, a loop with a half-twist. Cut this strip parallel to one edge about one-third of the way from the edge. Continue cutting until you get back to your starting point. Describe the result.

   ____

4. Prepare a loop with a full twist. Cut this strip parallel to one edge about one-third of the way from the edge. Continue cutting until you get back to your starting point. Describe the result.

   ____

   ____

5. Prepare a loop with three half-twists. Cut this strip parallel to one edge about one-half of the way from the edge. Continue cutting until you get back to your starting point. Describe the result.

   ____

Name ______________________ Date ______________ Class ______________

LESSON 7-8

Problem Solving

Symmetry

Complete the figure. A dashed line is a line of symmetry and a point is a center of rotation.

1.

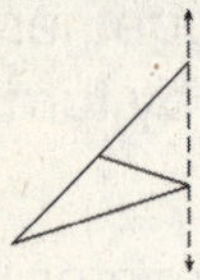

2.

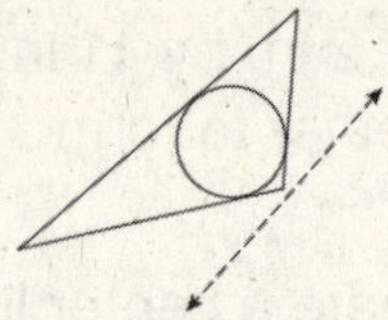

3. 2-fold symmetry

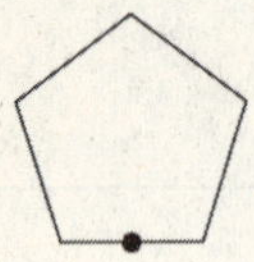

4. 4-fold symmetry

Use the flag of Switzerland to answer the questions.

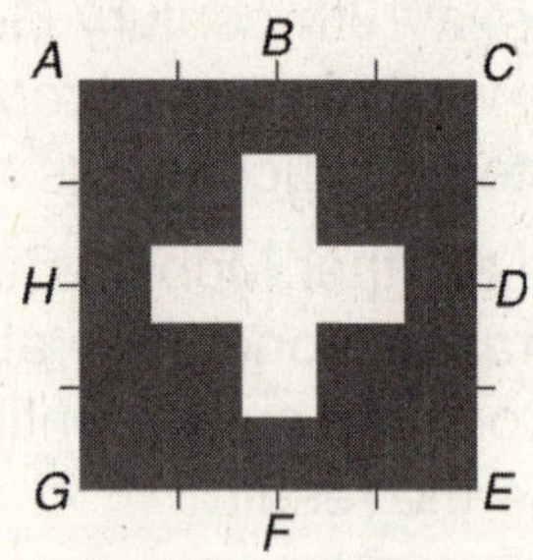

5. Which of the following would NOT be a line of symmetry?

   A $\overline{HD}$ C $\overline{AE}$

   B $\overline{BF}$ D $\overline{HB}$

6. How many lines of symmetry does the flag have?

   F 2 H 4

   G 6 J 8

7. How many folds of rotational symmetry does the flag have?

   A 0 C 2

   B 4 D 8

8. Which lists all lines of symmetry of the flag?

   F $\overline{AE}$, $\overline{GC}$

   G $\overline{HD}$, $\overline{BF}$

   H $\overline{HD}$, $\overline{BF}$, $\overline{AE}$, $\overline{GC}$

   J $\overline{HB}$, $\overline{DF}$, $\overline{AE}$, $\overline{GC}$

9. Which describes the center of rotation?

   A intersection of $\overline{BF}$ and $\overline{HD}$

   B intersection of $\overline{AE}$ and $\overline{HB}$

   C *A*

   D There is no center of rotation

Name ______________________ Date ______________ Class ______________

LESSON 7-8

Reading Strategies

Reading a Diagram

The star below turns around a center point. If a figure rotates onto itself before one complete rotation, the figure has **rotational symmetry**.

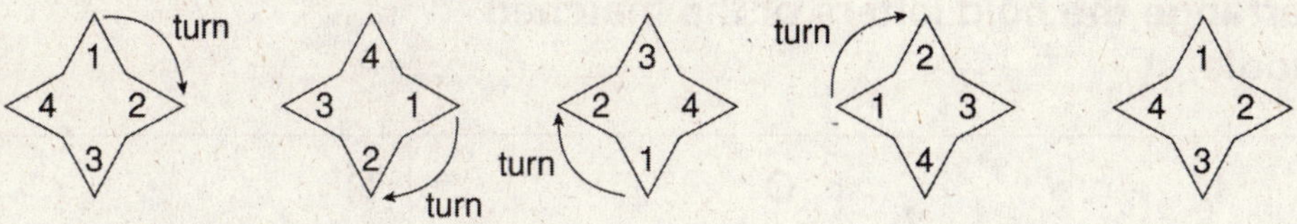

The star matches up with the original figure before it has made one complete turn. It has rotational symmetry

This right triangle does not have rotational symmetry. The rotated image matches up with the original triangle only after it has made one complete turn.

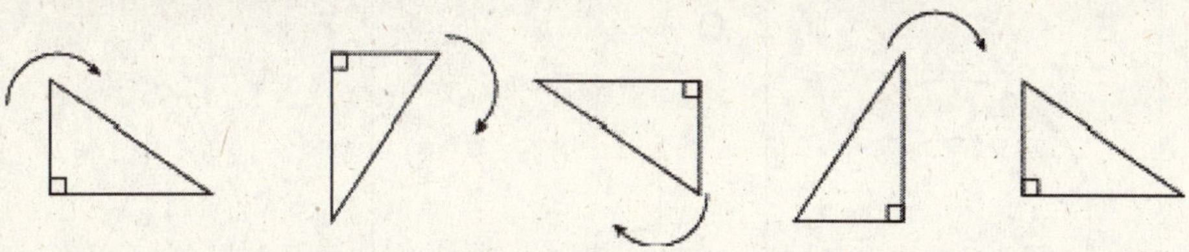

Do these figures have rotational symmetry? Trace each figure and rotate it to find out. Write *yes* or *no*.

1. 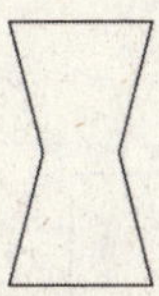 ______________________

2. 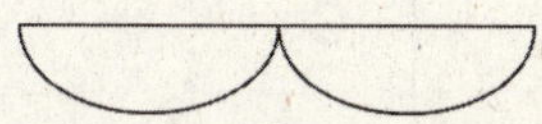 ______________________

A figure can be reflected over a line so that the two parts are congruent. That figure has line symmetry.

Line Symmetry

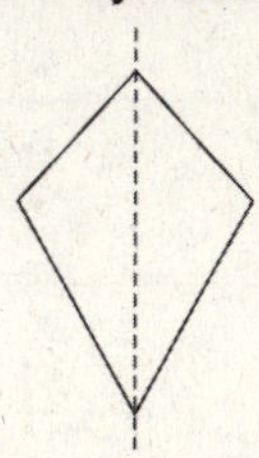

No Line Symmetry

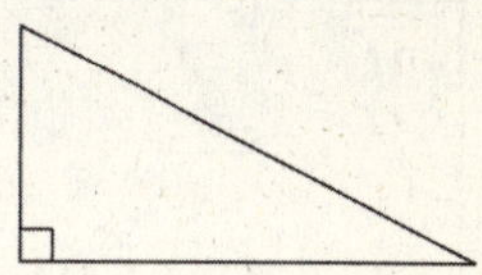

Do these figures have line symmetry? Trace each figure. Draw lines of symmetry if possible. Write *yes* or *no*.

3. 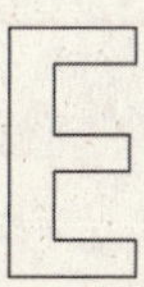

____

4. 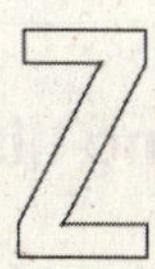

____

Name ____________________ Date ____________ Class ____________

LESSON 7-8

Puzzles, Twisters & Teasers

Concentrating on Symmetrical Figures

Imagine a game of concentration. Each box represents a card in the game. Choose two cards with figures that would have symmetry if they were put together without rotating the cards. Cross them out. Rearrange the bold letters of the matched cards to solve the riddle.

O 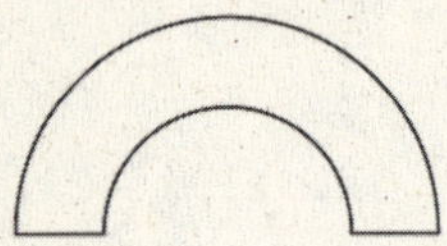	**I** 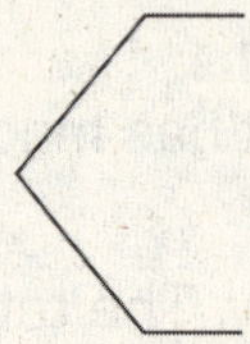	**O** 	**R** 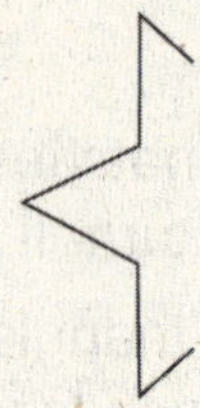
R 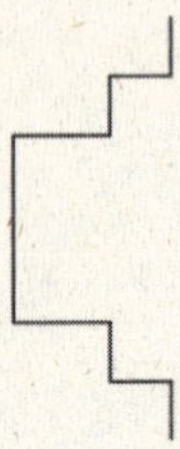	**E** 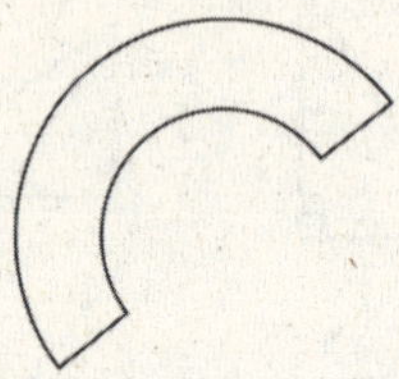	**O** 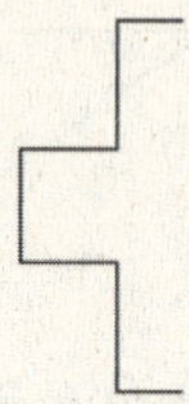	**T** 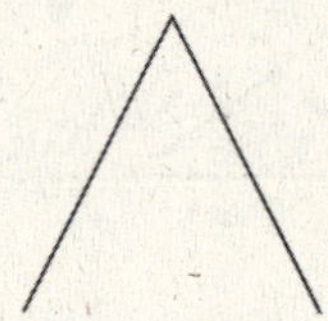
A 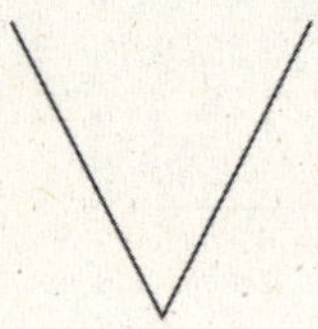	**C** 	**F** 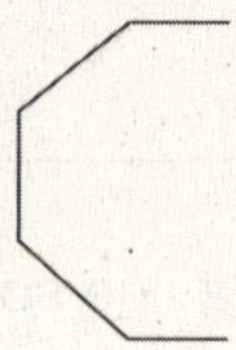	**X** 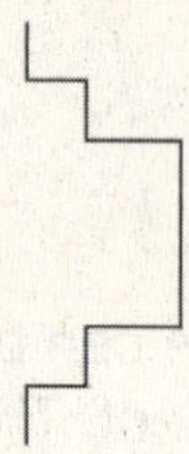
V 	**T** 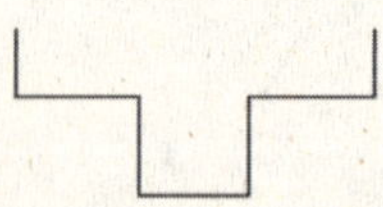	**D** 	**I** 

Who earns a living by driving his customers away?

A ___ ___ ___ ___   ___ ___ ___ ___ ___ ___

Answers

LESSON 7-1

Practice A

1. ∠*CED* or ∠*AEC*
2. ∠*AEB*, ∠*BEC* or ∠*DEF*
3. Possible answer: ∠*BED* and ∠*CEF*
4. ∠*AEB* and ∠*BEC*
5. Possible answer: ∠*AEC* and ∠*CED*, ∠*BEC* and ∠*CEF*
6. m∠3 = 120°
7. m∠4 = *x*°
8. m∠2 = 70°
9. m∠1 = 115°

Practice B

1. ∠*NXP*, ∠*MXR*, ∠*MXN*, or ∠*RXP*
2. ∠*MXS* and ∠*RXS*
3. ∠*NXS* and ∠*PXS*
4. ∠*RXS* and ∠*SXM*
5. Possible answer: ∠*NXP* and ∠*MXN*, ∠*RXM* and ∠*MXN*, ∠*RXP*, and ∠*PXN*
6. m∠4 = 110°
7. m∠3 = *n*°
8. m∠*ECD* = 150°

Practice C

1. true
2. false; Rays *YM* and *TM* make up line *YT*
3. true
4. true
5. false; ∠*YMX* and ∠*SMT* are vertical angles
6. true
7. false; If m∠*SMT* = 48°, then m∠*TMW* = 42°
8. true
9. true
10. true
11. 123°
12. 140°

Review for Mastery

1. right angle
2. acute angle
3. obtuse angle
4. straight angle
5. vertical angles
6. 90°; complementary angles
7. 180°; supplementary angles

Challenge

1.

Name of Coplanar Lines	Possible Number of Vertical Pairs
2	2
3	3
4	4
5	5

2. a. 4
   b. A,B; A,D; B,C; C,D
   c. m∠B,C,D = 90°

Problem Solving

1. Possible answer: ∠*DAB*
2. Possible answers: ∠*DAE*, ∠*EAB*
3. Possible answers: ∠*AED*, ∠*DAE*
4. Possible answers: ∠*DGI*, ∠*IGE*
5. C
6. F
7. C
8. H
9. C
10. J

Reading Strategies

1. ∠*XYZ*, ∠*ZYX*, or ∠*Y*
2. acute
3. 35°

Puzzles, Twisters & Teasers

C	O	M	P	L	E	M	E	N	T	A	R	Y	S
O	M	H	O	Y	U	P	U	I	O	N	W	E	U
N	O	L	I	T	Y	L	T	R	Y	G	R	T	P
G	I	U	N	R	T	A	W	E	R	L	U	I	P
R	A	Y	T	Q	W	N	P	L	M	E	Y	H	L
U	W	D	V	G	T	E	Q	W	E	R	T	Y	E
E	E	D	C	V	F	R	T	G	B	R	H	Y	M
N	O	I	U	Y	T	R	F	V	B	A	J	K	E
T	I	S	E	G	M	E	N	T	U	Y	T	F	N
I	M	L	P	O	K	N	B	J	L	U	H	V	T
M	W	D	C	F	R	T	G	H	N	I	J	U	A
Q	V	E	R	T	I	C	A	L	D	C	N	H	R
E	P	A	N	T	S	U	J	M	K	I	O	E	Y

P A N T S

LESSON 7-2

Practice A

1. ∠1 ≅ ∠3 ≅ ∠5 ≅ ∠7 and ∠2 ≅ ∠4 ≅ ∠6 ≅ ∠8
2. 50°
3. 130°
4. 50°
5. 130°
6. 50°
7. 130°
8. 45°
9. 135°
10. 45°
11. 45°
12. 135°
13. 135°
14. ∠3, ∠5, ∠7
15. ∠4, ∠6, ∠8
16. Possible answer: ∠1 + ∠8, ∠2 + ∠7, ∠3 + ∠6, or ∠4 + ∠5
17. line *c*

Practice B

1. ∠1 ≅ ∠4 ≅ ∠5 ≅ ∠8 and ∠2 ≅ ∠3 ≅ ∠6 ≅ ∠7
2. 38°; ∠1 and the 142° angle are supplementary angles
3. 142°; ∠2 and the 142° angle are vertical angles
4. 38°; ∠1 and ∠5 are corresponding angles
5. 142°; ∠5 and ∠6 are supplementary angles
6. 142°; ∠8 and the 142° angle are alternate exterior angles
7. 38°; ∠6 and ∠7 are supplementary angles
8. 43°; ∠6 and the 137° are supplementary and ∠6 and ∠2 are alternate interior angles
9. 137°; ∠5 and the 137° angle are vertical angles
10. 43°; ∠5 and ∠6 are supplementary angles
11. 43°; ∠7 and the 137° angle are supplementary angles
12. 43°; ∠4 and ∠6 are corresponding angles
13. 137°; ∠3 and ∠5 are alternate interior angles
14. ∠4, ∠6, ∠8
15. line *t*

Practice C

1. ∠3 ≅ ∠5 ≅ ∠7
2. ∠4 ≅ ∠6 ≅ ∠8
3. Possible answer: ∠1 + ∠8, ∠2 + ∠7, ∠3 + ∠6, or ∠4 + ∠5
4. line *c*
5. 49°
6. 123°
7. 53°
8. 90°
9. Possible answer:

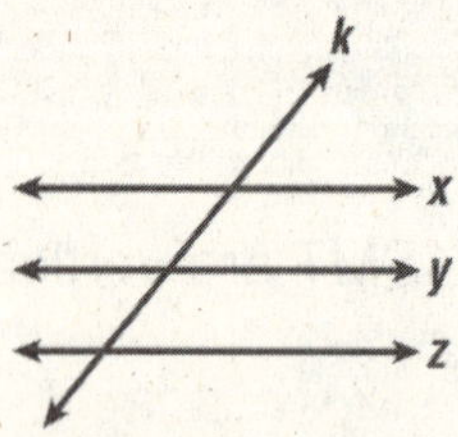

10. Possible answer:

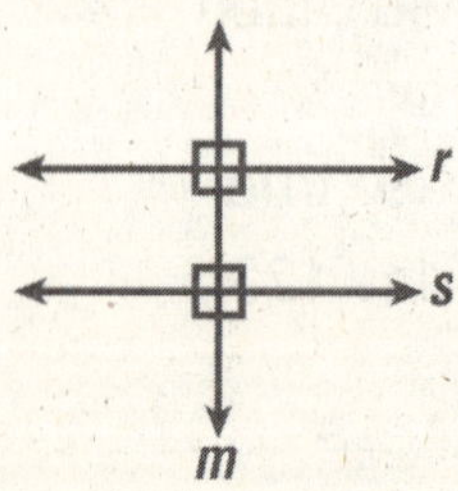

11. Possible answer:

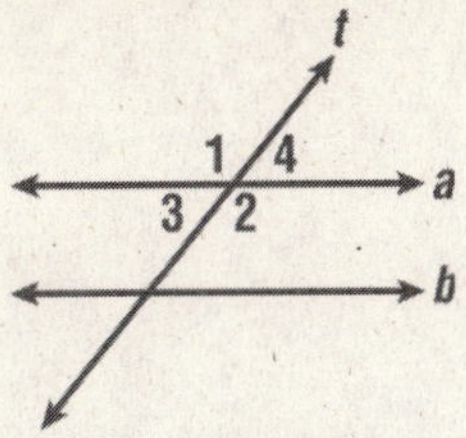

Review for Mastery

1. $\angle 3$, $\angle 5$, $\angle 7$
2. $\angle c$, $\angle e$, $\angle g$
3. $\angle y$, $\angle t$, $\angle r$
4. 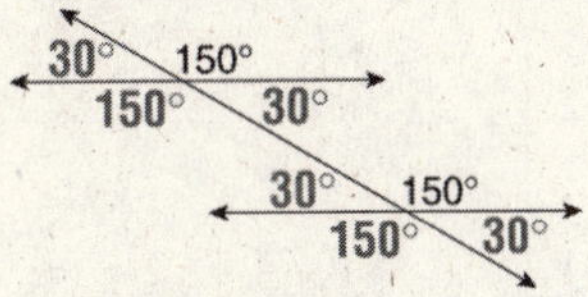

5. 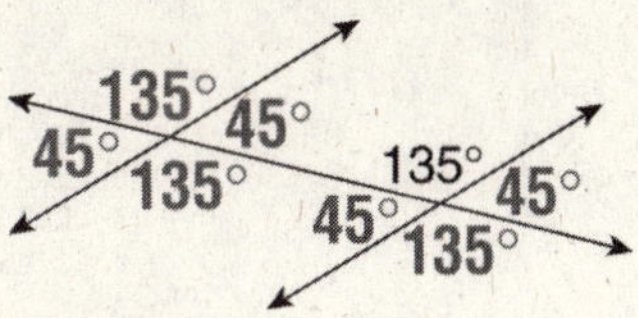

6. 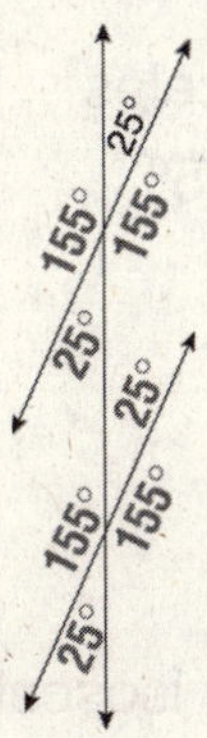

Challenge

1. $\angle 3$, $\angle 4$, $\angle 5$, $\angle 6$
2. $\angle 3$ and $\angle 5$; $\angle 4$ and $\angle 6$
3. $\angle 3 \cong \angle 5$
4. $\angle 4 \cong \angle 6$
5. Alternate interior angles of parallel lines are equal in measure, or congruent.
6. 48°
7. 35° + 45° = 80°
8. 70° – 30° = 40°

Problem Solving

1. $\angle 3$, $\angle 5$, $\angle 7$, $\angle 9$
2. $\angle 4$, $\angle 6$, $\angle 8$, $\angle 10$
3. Possible answer: $\angle 1$, $\angle 2$
4. $m\angle 3 = m\angle 5 = m\angle 7 = m\angle 9 = 75°$, $m\angle 2 = m\angle 4 = m\angle 6 = m\angle 8 = m\angle 10 = 105°$
5. Possible answer: $\angle 2$, $\angle 8$
6. Possible answer: $\overline{AB}$, $\overline{CD}$, $\overline{EF}$
7. B
8. H
9. B
10. H
11. B
12. H
13. B

Reading Strategies

1. Possible answer. $\overline{EF}$ is parallel to $\overline{GH}$.
2. $\overline{EF} \parallel \overline{GH}$
3. $\overline{QR} \perp \overline{TR}$
4. Possible answer: $\overline{QS} \perp \overline{ST}$
5. the square in the corner

Puzzles, Twisters & Teasers

1. R
2. E
3. C
4. C
5. I
6. N

R E C Y C L I N G

LESSON 7-3

Practice A

1. acute, isosceles
2. right, scalene
3. obtuse, isosceles
4. 65°
5. 40°
6. 30°
7. 30°
8. 42°
9. 25°
10. 70°
11. 40°
12. 65°
13. Possible answer: 36°, 108°, 36°

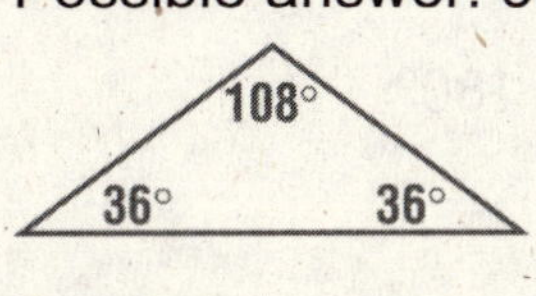

Practice B

1. 41°
2. 91°
3. 74°
4. 68°
5. 65°
6. 53°
7. 58°
8. 60°

9. 62°

10. Possible answer: 30°, 60°, 90°

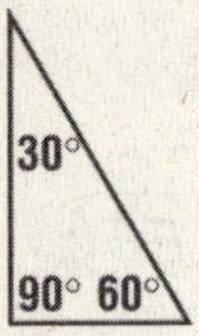

11. Yes; $6 + 8 > 13$

12. No; $8 + 2 \not> 15$

13. Yes; $9 + 14 > 22$

Practice C

1. 59°
2. 46°
3. 60°
4. 31°
5. 57°
6. 30°
7. first ∠ = 42°; second ∠ = 46°; third ∠ = 92°
8. sometimes
9. never
10. sometimes
11. always

Possible answer:

12. 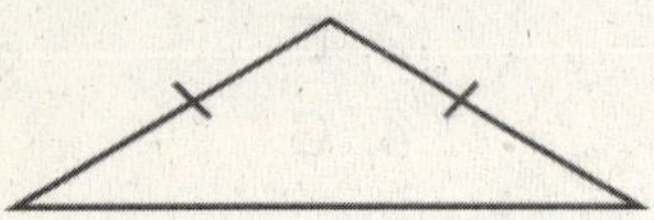

13. 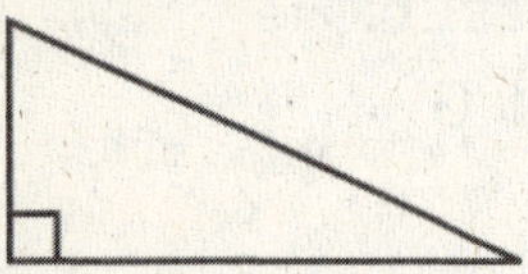

14. 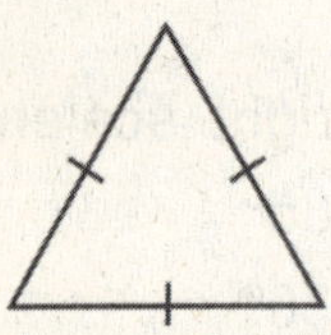

Review for Mastery

1. 58°; 42°; 180°
   100°; 180°
   80°
2. 32°; 90°; 180°
   122°; 180°
   58°
3. $x° + 35° + 35° = 180°$
   $x° + 70° = 180°$
   110°
4. $x° + 20° + 122° = 180°$
   $x° + 142° = 180°$
   38°

Challenge

1. 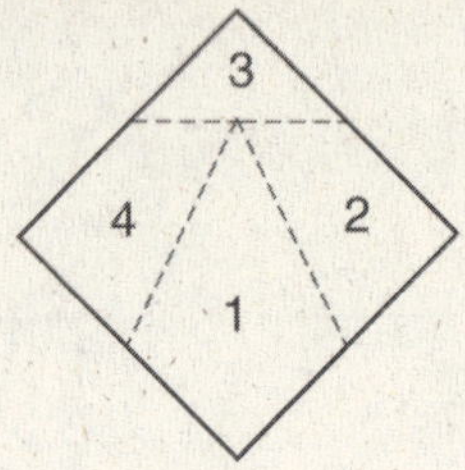

2. 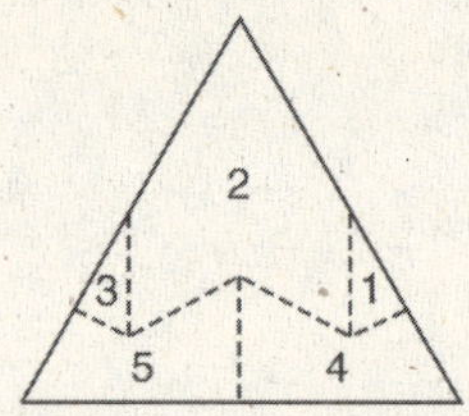

3. 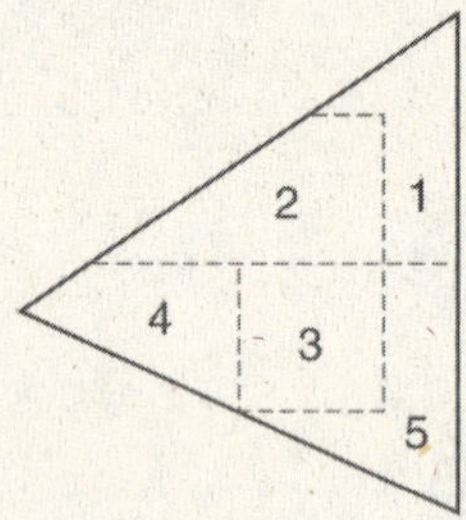

Problem Solving

1. right
2. isosceles
3. $x = 45°$
4. $y = 45°$
5. B
6. H
7. C
8. F
9. C
10. H

Reading Strategies

1. equilateral, acute
2. right, isosceles

Puzzles, Twisters & Teasers

Across

1. RIGHT
5. ACUTE
6. THEOREM
7. SCALENE

Down

2. ISOSCELES
3. EQUILATERAL
4. OBTUSE

LESSON 7-4

Practice A

1. pentagon
2. octagon
3. hexagon
4. 360°
5. 180°
6. 540°

7. 360°
8. 720°
9. 1080°
10. 60°
11. 144°
12. 108°
13. 135°
14. 90°
15. 120°
16. quadrilateral, parallelogram, rectangle
17. quadrilateral, parallelogram
18. quadrilateral, parallelogram, rectangle, rhombus, square

Practice B

1. 360°
2. 720°
3. 900°
4. 180°
5. 1080°
6. 540°
7. 120°
8. 90°
9. 135°
10. 108°
11. 60°
12. 144°
13. quadrilateral, parallelogram, rhombus
14. quadrilateral
15. quadrilateral, trapezoid

Practice C

1. 3960°; 165°
2. 2520°; 157.5°
3. 6120°; 170°
4. 55°
5. 40°
6. 64°
7. octagon
8. nonagon
9. heptagon
10. decagon
11. octagon-eight sides

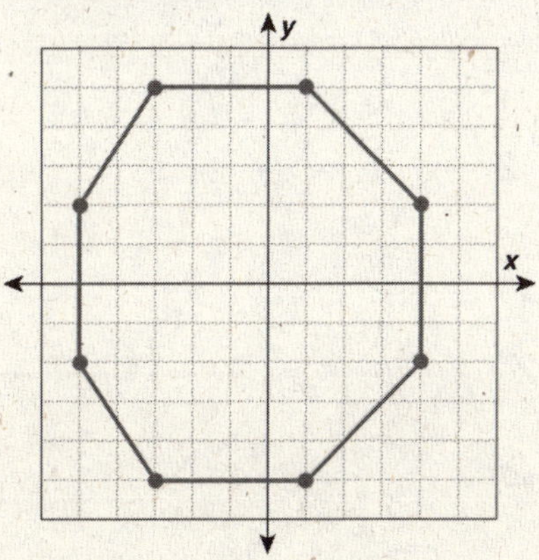

12. pentagon-five sides

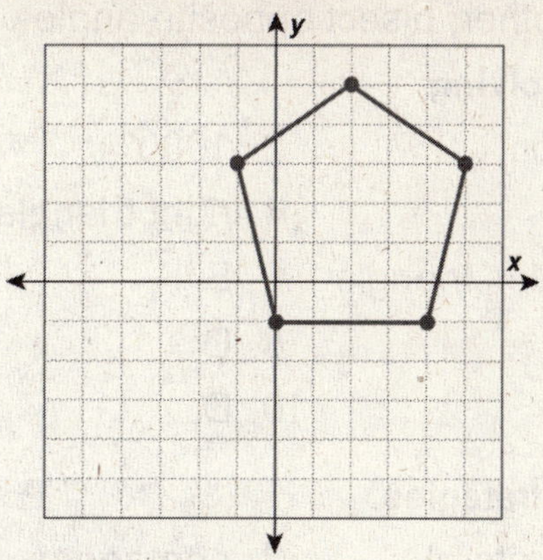

Review for Mastery

1. 4
   4; 2
   2
   360°
2. 6
   6; 4
   4
   720°
3. 8
   8; 6
   6
   1080°
   $\frac{1080°}{8}$
   135°
4. 8
   10; 8
   8
   1440°
   $\frac{1440°}{10}$
   144°
5. quadrilateral
   trapezoid
6. quadrilateral
   parallelogram
   rhombus
7. quadrilateral
   parallelogram
   rectangle

Challenge

1. a. no
   b. yes; The diagonals of a parallelogram are not congruent but they do bisect each other.
2. a. Diagonals bisect each other.
   b. Diagonals are congruent.
3. a. Diagonals bisect each other.
   b. no
   c. Diagonals are perpendicular to each other.
   d. Diagonals bisect the opposite angles.

4. congruent; perpendicular bisector of each other; bisect opposite angles

Problem Solving

1. octagon
2. 1080º
3. 135º
4. right triangle
5. isosceles triangle
6. B
7. F
8. D
9. H
10. D

Reading Strategies

1. quadrilateral
2. pentagon
3. hexagon
4. octagon

Puzzles, Twisters & Teasers

1. P E N T A G O N
2. O C T A G O N
3. T R I A N G L E
4. H E P T A G O N
5. Q U A D R I L A T E R A L
6. H E X A G O N

A H A R D D R I V E

LESSON 7-5

Practice A

1.

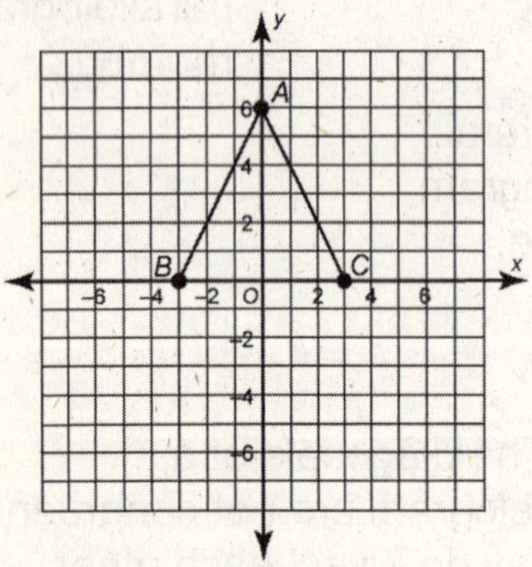

isosceles triangle

2.

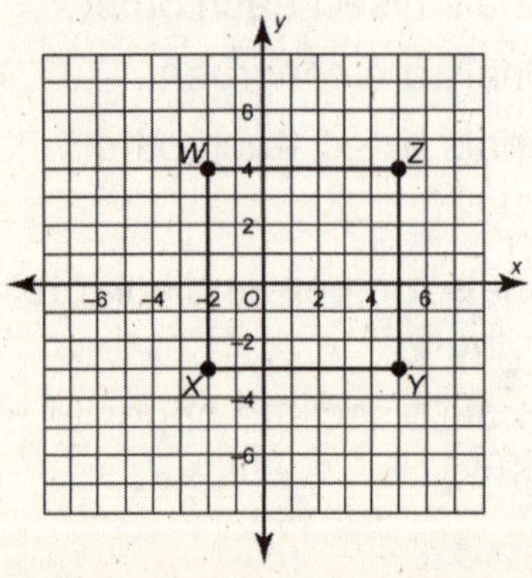

square

3. (–3, 0) or (–3, –8)
4. (–2, 3)
5. $(-3\frac{1}{2}, 1)$
6. $(1, -\frac{1}{2})$
7. $(-1\frac{1}{2}, -5\frac{1}{2})$
8. (–2, 1)
9. (0, 5)
10. (4, –1)

Practice B

1.

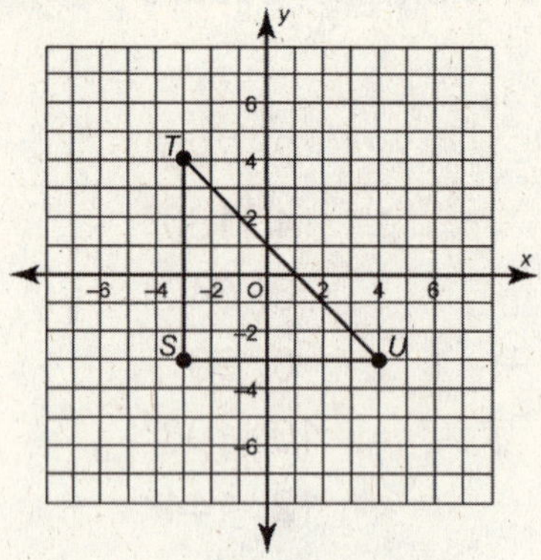

right isosceles triangle

2.

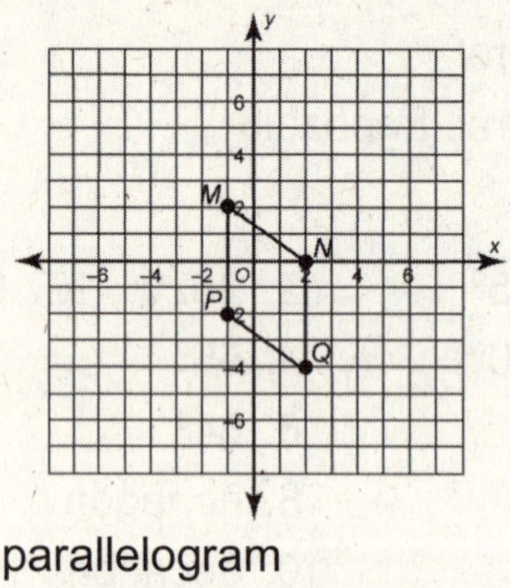

parallelogram

3. (–2, 6) or (–2, –4)
4. (4, –5)
5. $(2\frac{1}{2}, 3\frac{1}{2})$
6. (–3, 5)
7. $(-1, 7\frac{1}{2})$
8. $(1\frac{1}{2}, -2)$
9. $(-1\frac{1}{2}, -7\frac{1}{2})$

Practice C

1. 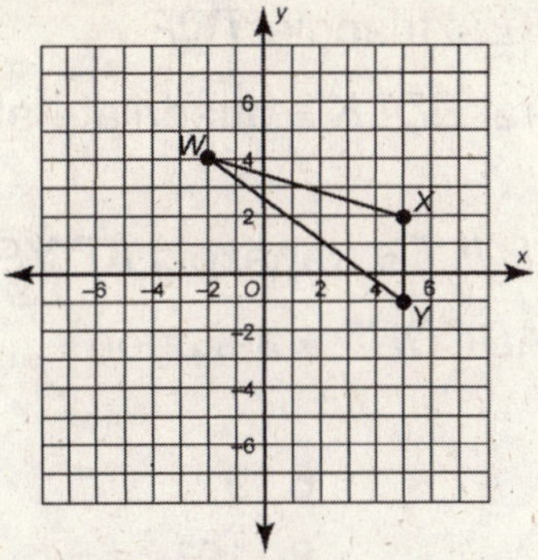

scalene triangle

2. 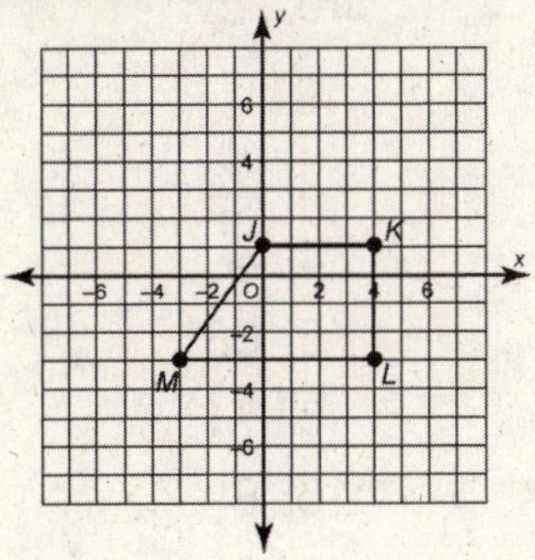

trapezoid

3. (5, –2) 4. E(2, 0)

5. G(3, 4) 6. X(3, 5)

7. A(–2, –1) 8. $(\frac{1}{2}, -\frac{1}{2})$

9. $(-1, -7\frac{1}{2})$ 10. $(4, -3\frac{1}{2})$

Review for Mastery

1. 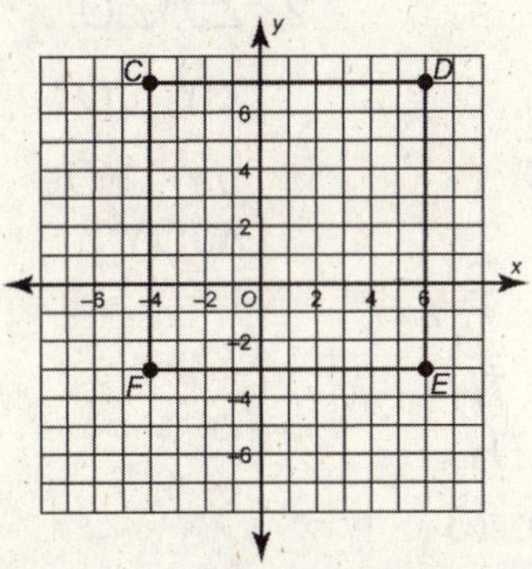

4 right angles, 2 pairs of perpendicular lines, 2 pairs of parallel lines; square

2. 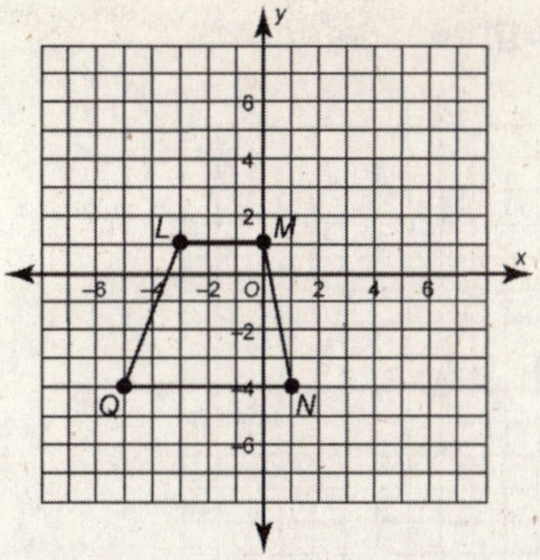

1 pair of parallel lines; trapezoid

3. (0, 1) 4. $(-\frac{1}{2}, -\frac{1}{2})$

5. (3, 3) 6. (1, 4)

7. $(-2, 5\frac{1}{2})$

Challenge

1. (4, –5) 2. (–8, –5)

3. (–4, 11) 4. (22, 5)

5. $(-3\frac{1}{2}, -1), (3\frac{1}{2}, -1), (0, -7)$

6. 4 7. 16

Problem Solving

1. 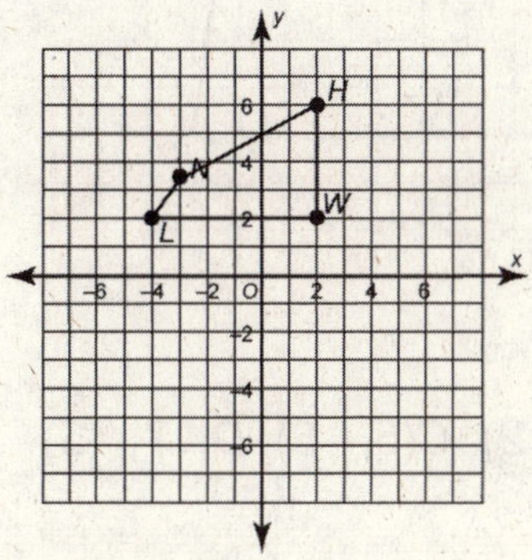

quadrilateral

2. $(5\frac{1}{2}, 2)$ or $(-1\frac{1}{2}, 2)$

3. (–1, 4) 4. D

5. G 6. B

7. H

Reading Strategies

1.

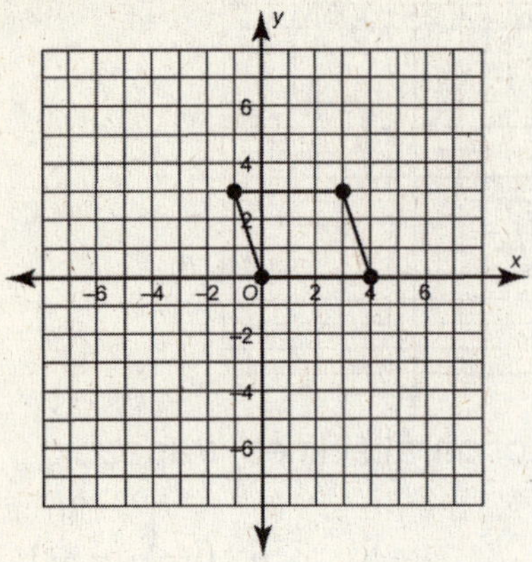

2. 2
3. 0
4. 2
5. parallelogram
6. (–2, 4)
7. $(4\frac{1}{2}, 1\frac{1}{2})$

Puzzles, Twisters & Teasers

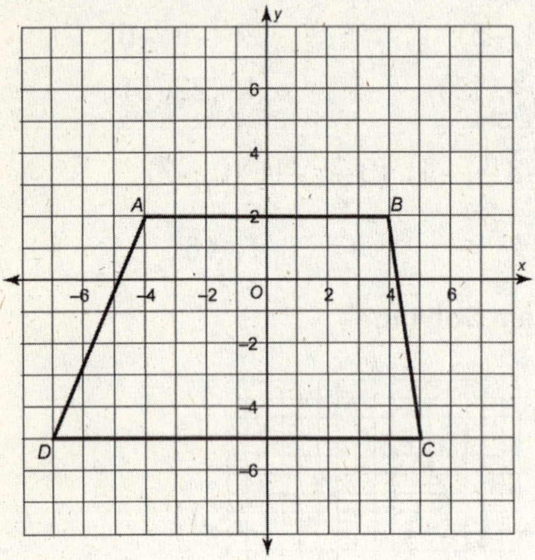

trapezoid

A(–4, 2); *B*(4, 2); *C*(5, –5), *D*(–7, –5)

LESSON 7-6

Practice A

1. B
2. D
3. A
4. C
5. triangle *ABC* ≅ triangle *DEF*
6. parallelogram *ABCD* ≅ parallelogram *TRNM*
7. 10
8. 55º

Practice B

1. triangle *JKL* ≅ triangle *TQP*
2. quadrilateral *BDLK* ≅ quadrilateral *JHRP*
3. trapezoid *BJLT* ≅ trapezoid *KDYP*
4. hexagon *ABCDEF* ≅ hexagon *VTZYXW*
5. 4
6. 9
7. 10
8. 10º
9. 105º
10. 15º

Practice C

1. 4
2. 22
3. 9
4. 4
5. 45º
6. 68º
7. 87º
8. 71º
9. 1
10. 8
11. 137º
12. 6
13. 56º
14. 12
15. 4
16. 3
17. 32º
18. 11º
19. 152º
20. 128º
21. 4
22. 5
23. 132º
24. 298º

Review for Mastery

1. $\angle A'$; $\angle B'$; $\angle C'$; $\overline{A'B'}$; $\overline{B'C'}$; $\overline{A'C'}$; $\Delta A'B'C'$
2. $\angle J$; $\angle O$; $\angle Y$; $\overline{JO}$; $\overline{JY}$; $\overline{YO}$; ΔJOY
3. $\angle M$; $\angle A$; $\angle T$; $\angle H$; $\overline{MA}$; $\overline{AT}$; $\overline{TH}$; $\overline{HM}$; quad. *MATH*
4. 10; +3; +3; 13
5. $\angle S$; 120; $\frac{120}{2}$; 2; 60
6. $\angle F$; 150; 2; 150; 2; 300
7. $\overline{FE}$; 14; –2; –2; 12; 3; $\frac{12}{3}$; 4

Challenge

1.

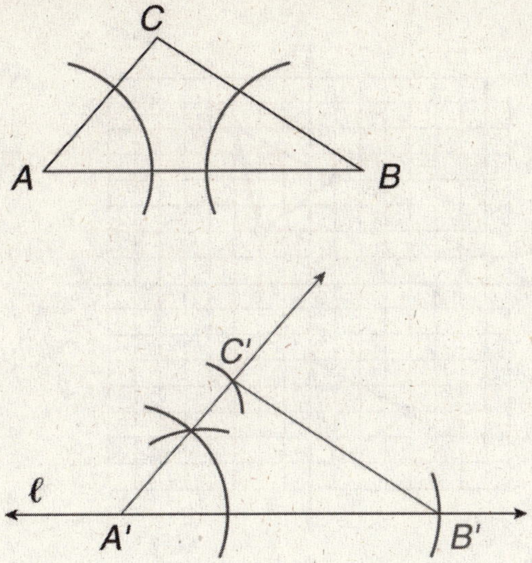

2.

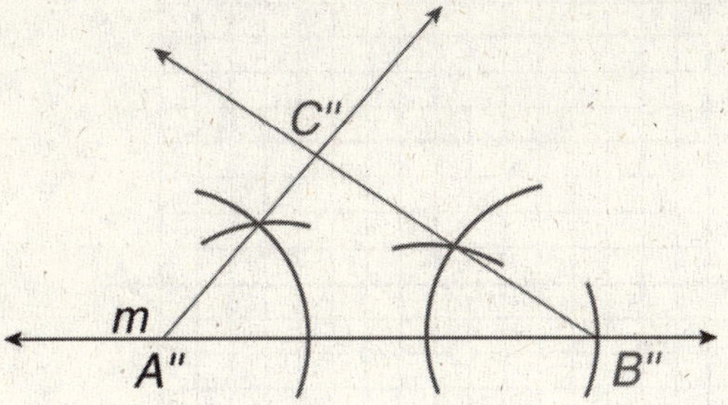

Problem Solving

1. 45º
2. 4 inches
3. 6 inches
4. 90º
5. 4 inches
6. B
7. G
8. D
9. J
10. C
11. J
12. C

Reading Strategies

1. They have the same size and shape.
2. no
3. no
4. no
5. yes
6. yes

Puzzles, Twisters & Teasers

1. K
2. F
3. E
4. L
5. C
6. A
7. O

C L O C K   F A C E

LESSON 7-7

Practice A

1.

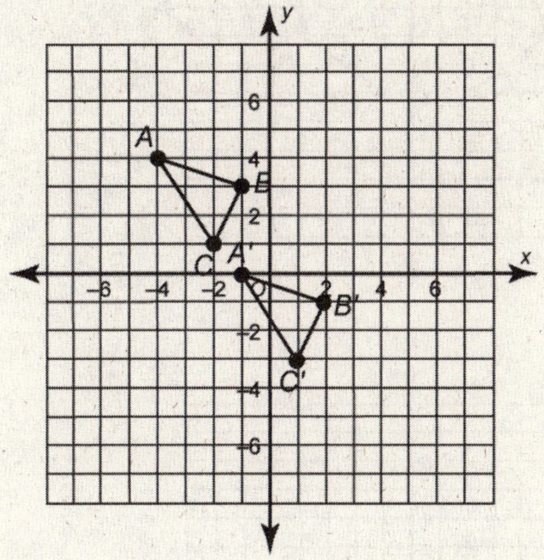

2.

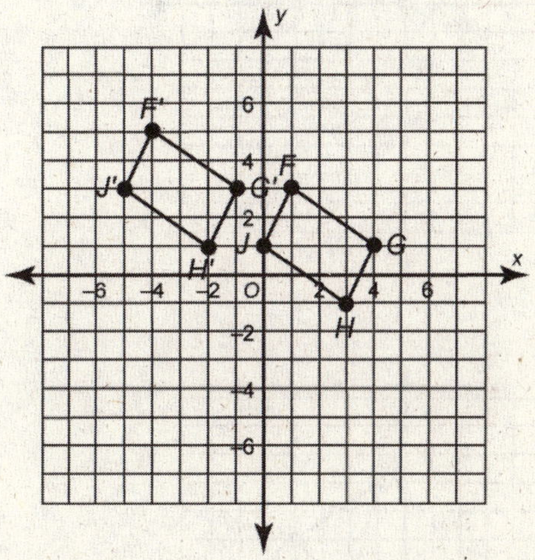

3.

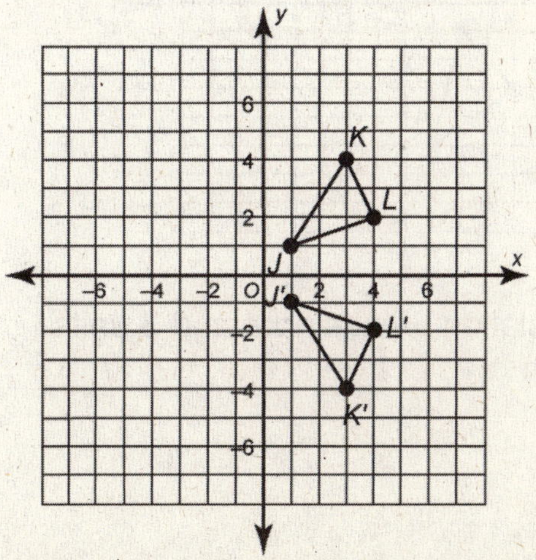

4.

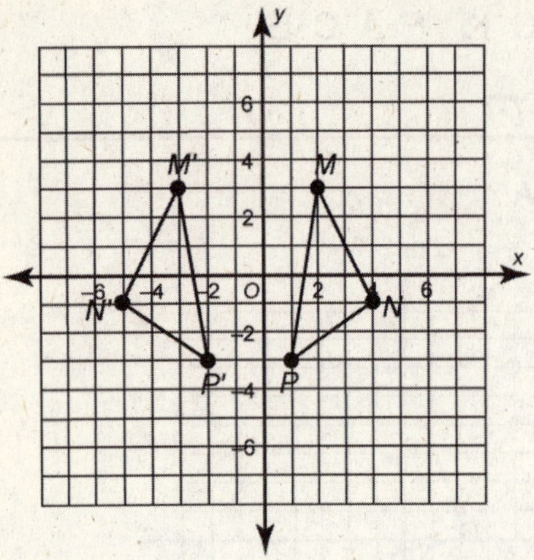

5.

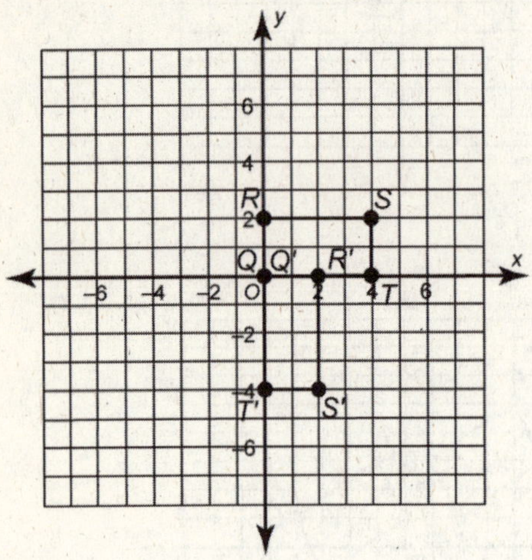

6.

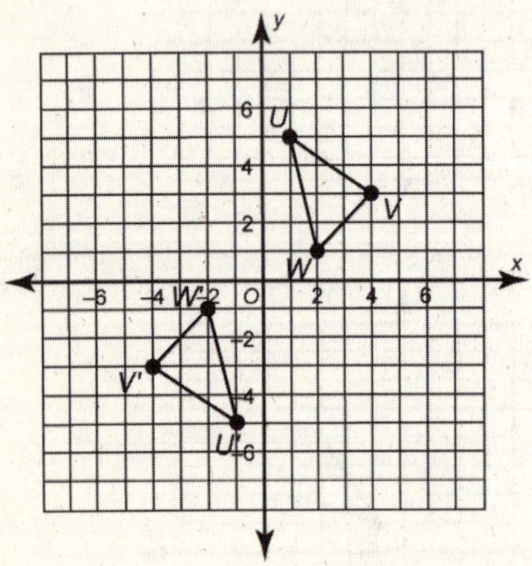

7. translation 6 units right and 3 units down

Practice B

1.

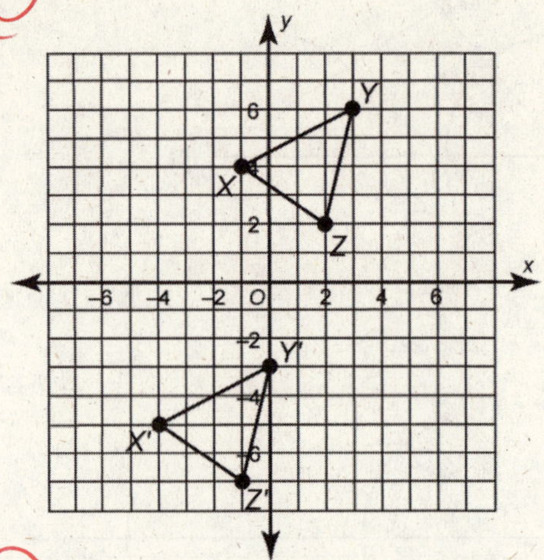

2.

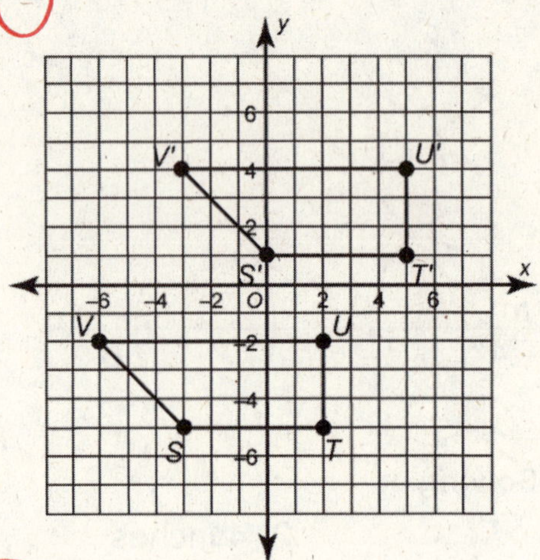

3.

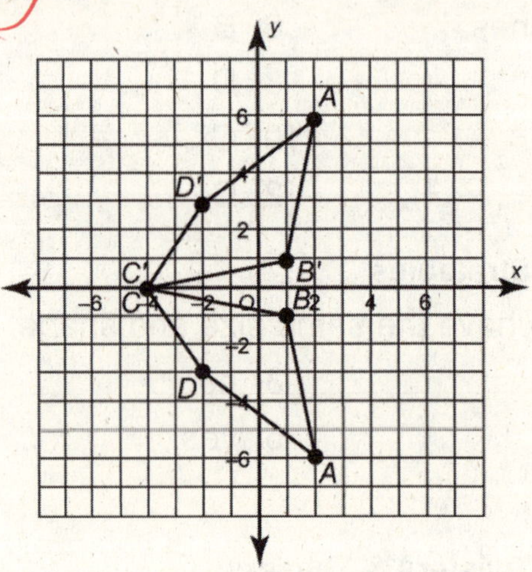

4.

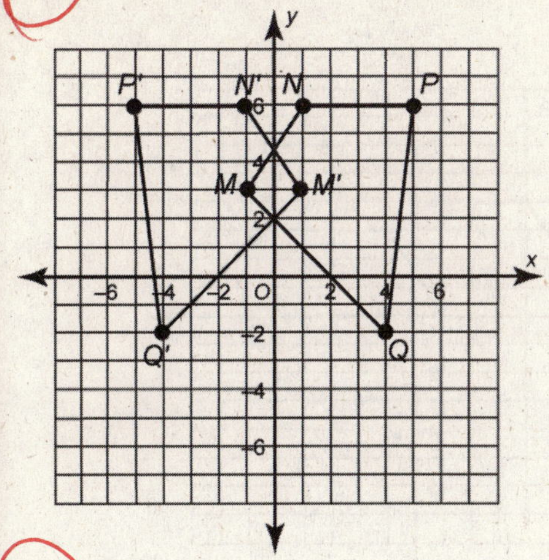

5.

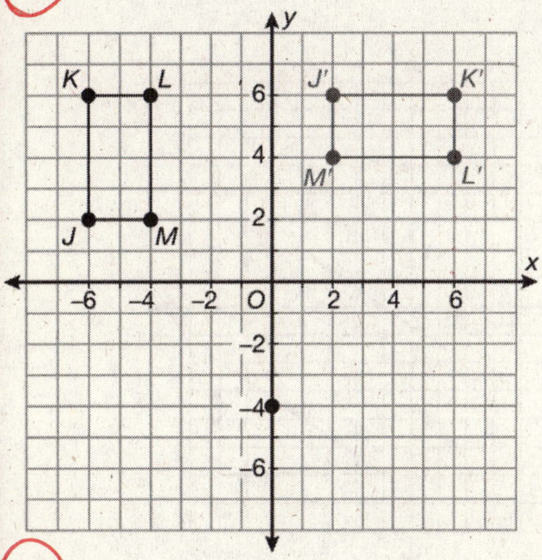

6.

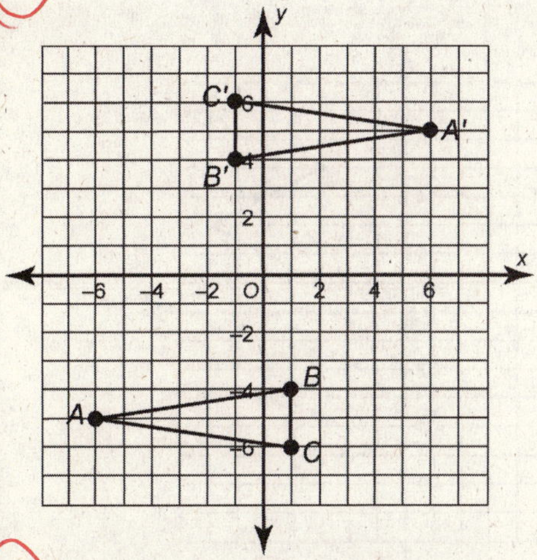

7. reflection across the *y*-axis

Practice C

1.

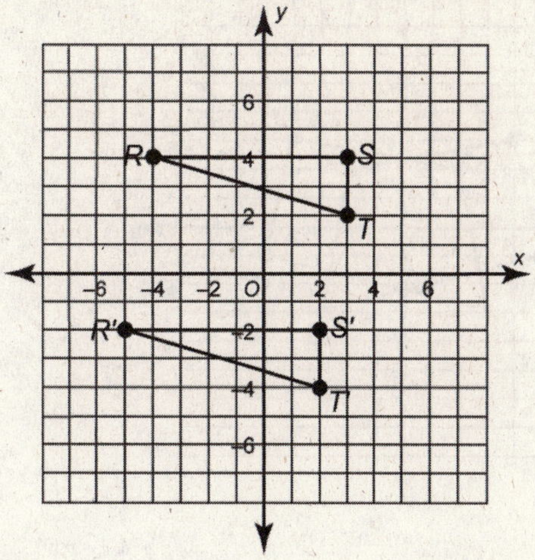

2.

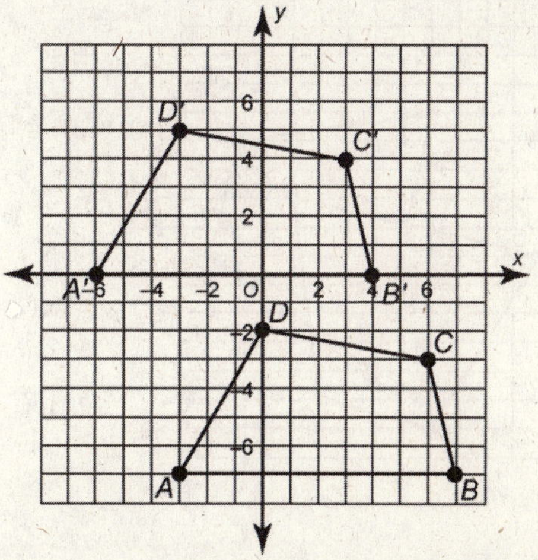

3.

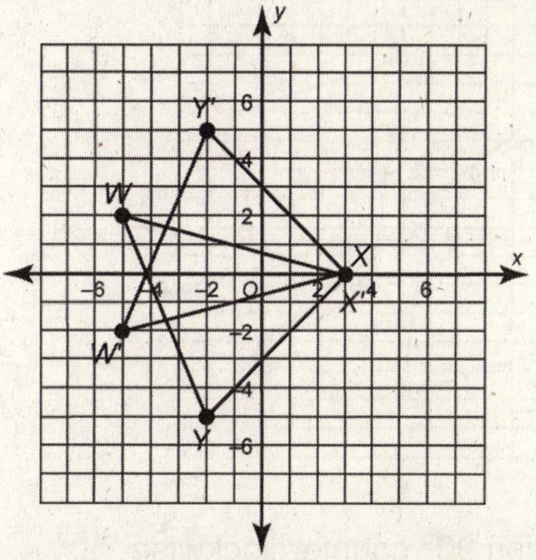

4.

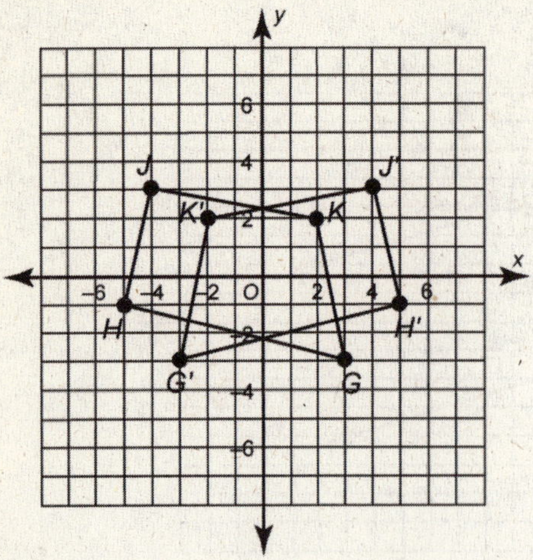

5.

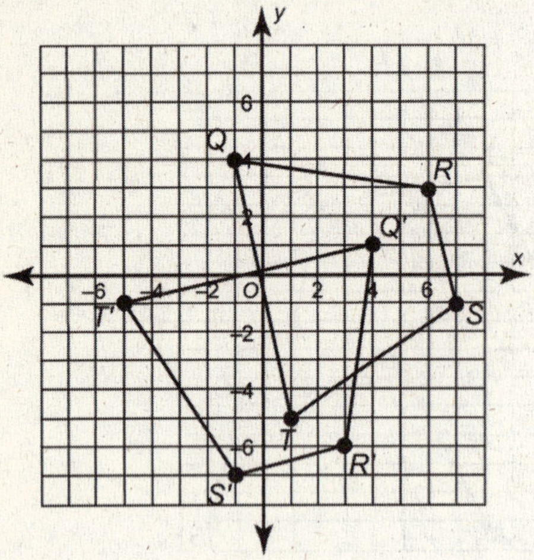

6.

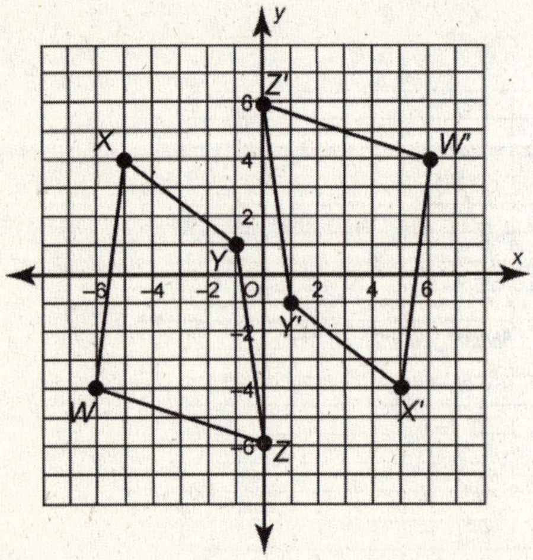

7. rotation 90° counterclockwise

Review for Mastery

1.

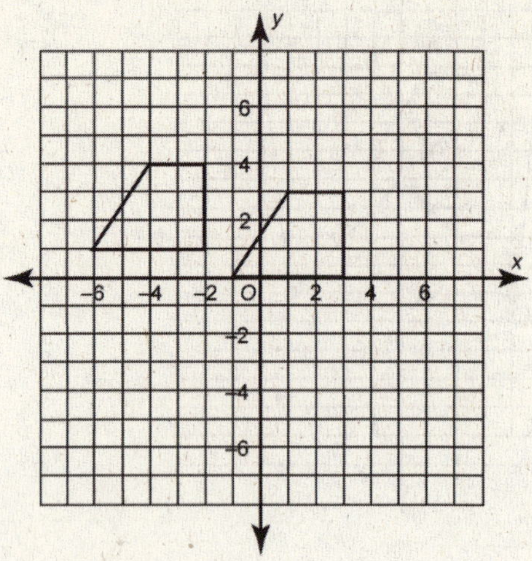

2.

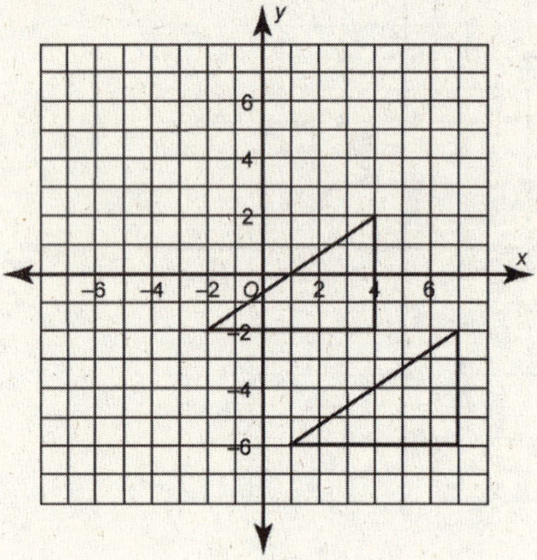

3.

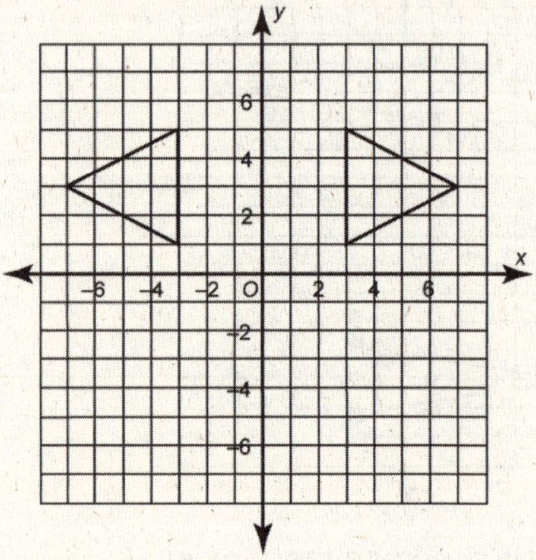

4.

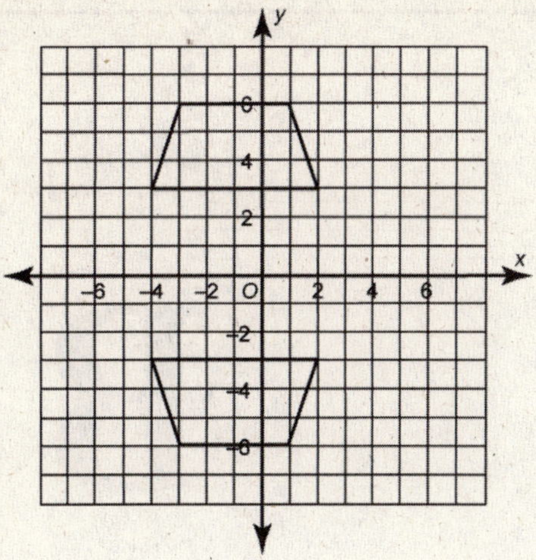

5.

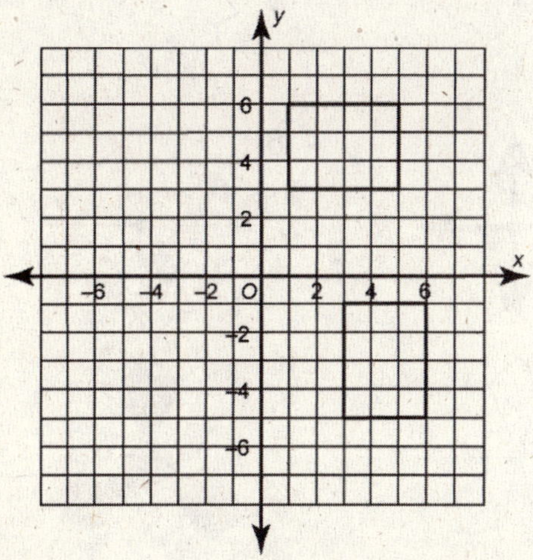

6.

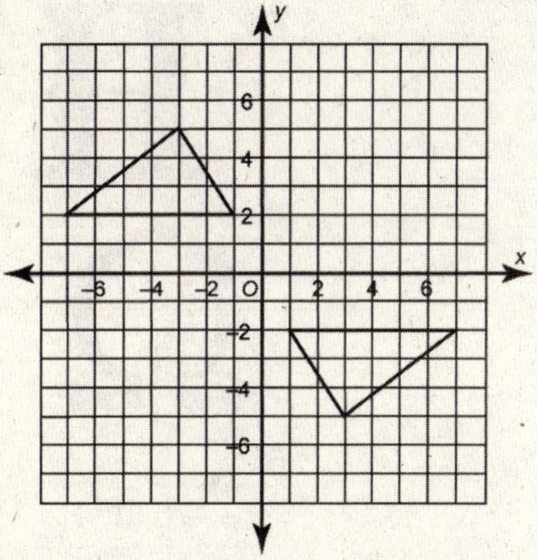

Challenge

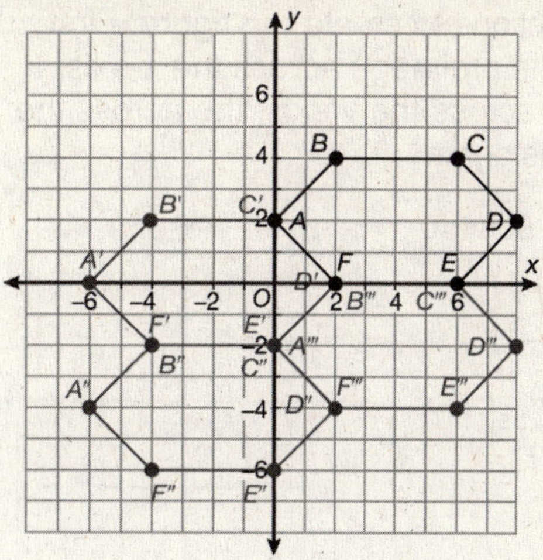

1. *A*′(–6, 0), *B*′(–4, 2), *C*′(0, 2), *D*′(2, 0), *E*′(0, –2), *F*′(–4, –2)
2. *A*″(–6, –4), *B*″(–4, –2), *C*″(0, –2), *D*″(2, –4), *E*″(0, –6), *F*″(–4, –6)
3. *A*‴(0, –2), *B*‴(2, 0), *C*‴(6, 0), *D*‴(8, –2), *E*‴(6, –4), *F*‴(2, –4)
4. Yes, the pattern is congruent. The figures cover the plane without overlapping and without any gaps.

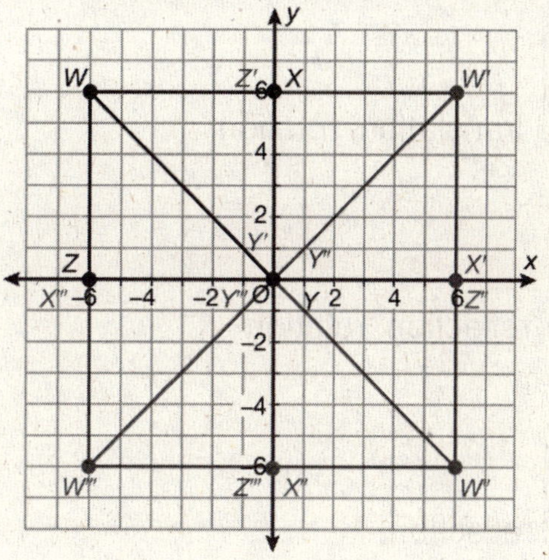

5. *W*′(6, 6), *X*′(6, 0), *Y*′(0, 0), *Z*′(0, 6)
6. *W*″(6, –6), *X*″(0, –6), *Y*″(0, 0), *Z*″(6, 0)
7. *W*‴(–6, –6), *X*‴(–6, 0), *Y*‴(0, 0), *Z*‴(0, –6)
8. Yes, the figures cover the plane without overlapping and without any gaps.

9. Yes, you could use a series of reflections to create the figure. First flip the preimage across the *x*-axis, then across the *y*-axis, the across the *x*-axis again.

Problem Solving

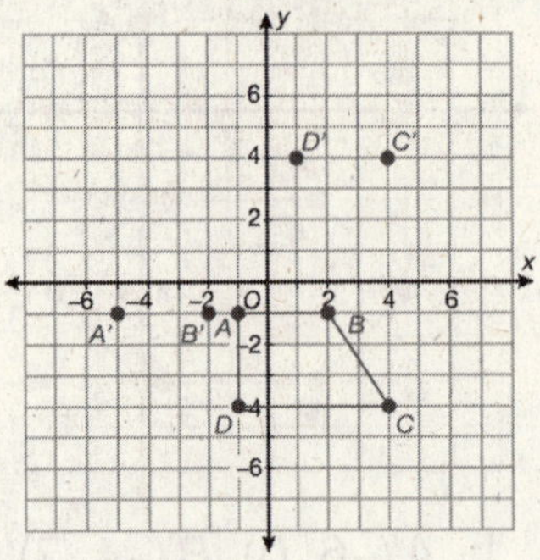

1. (–5, –1)
2. (–2, –1)
3. (4, 4)
4. (1, 4)
5. B
6. F
7. D
8. G

Reading Strategies

1. rotation
2. reflection
3. translation

Puzzles, Twisters & Teasers

1–4. Possible answers are given.

1. 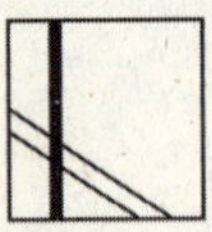 translation, rotation

2. 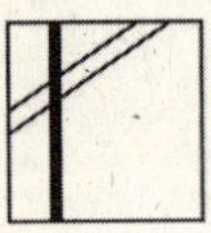 reflection, rotation

3. 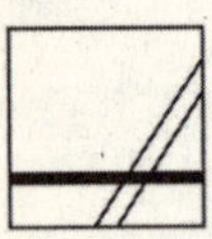 reflection, 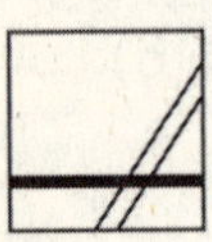 translation, rotation

4. 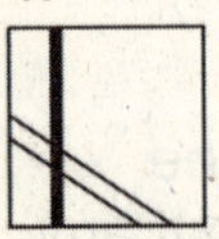 reflection, 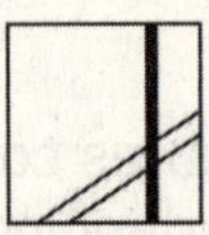 reflection, 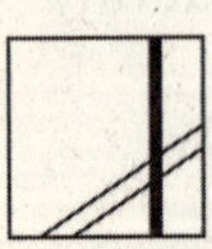 translation, rotation

LESSON 7-8

Practice A

1. 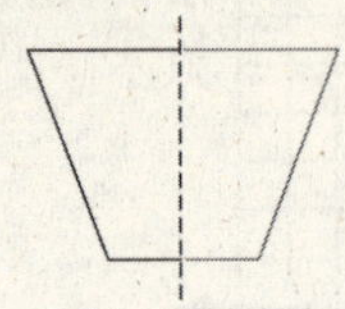
2. 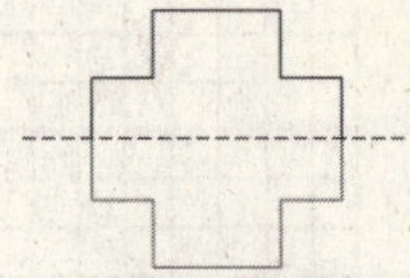
3. 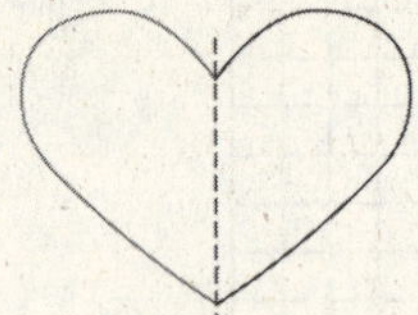
4. 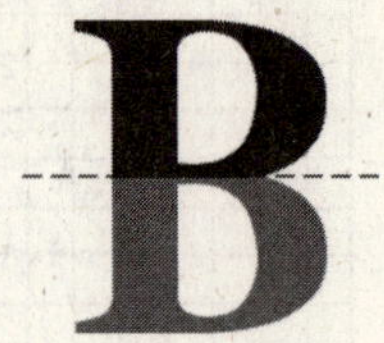
5. 
6. 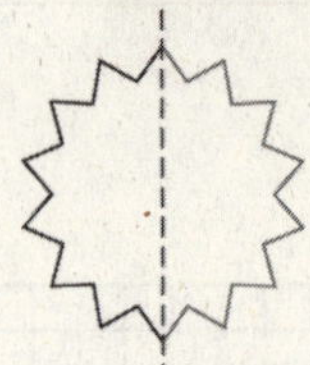
7. 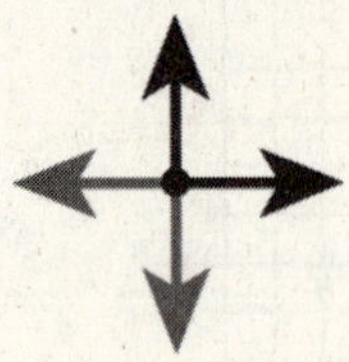
8. 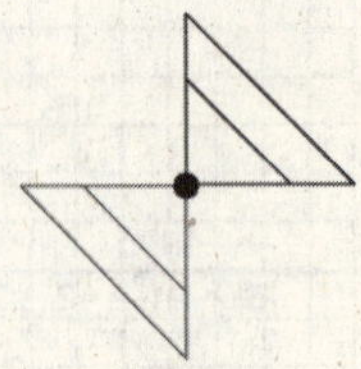
9. 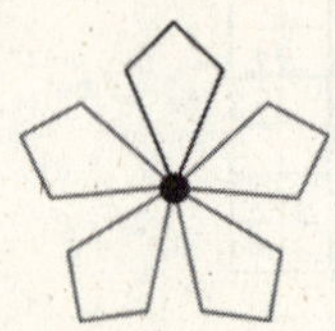
10. 

Practice B

1. 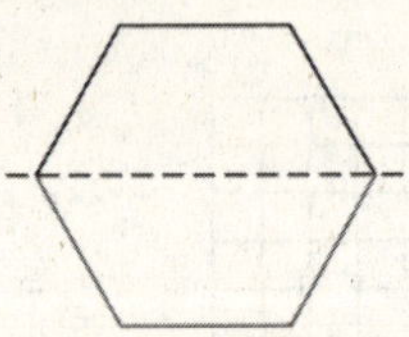
2. 
3. 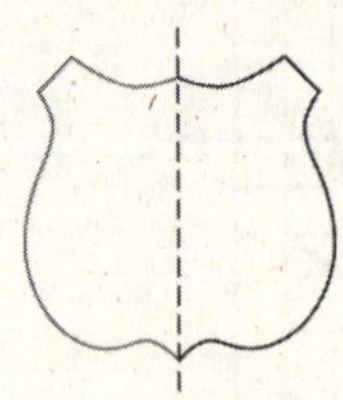
4. 
5. 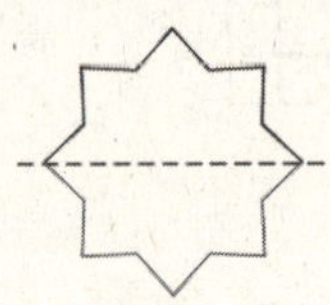
6. 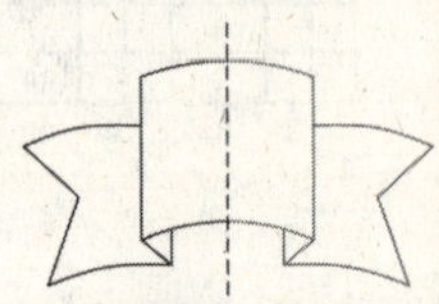
7. 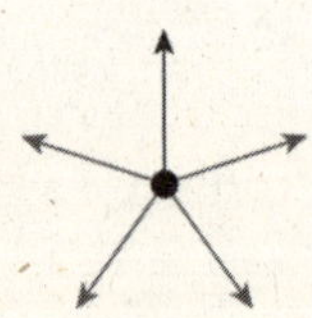
8. 

9. 

10. 

Practice C

1. 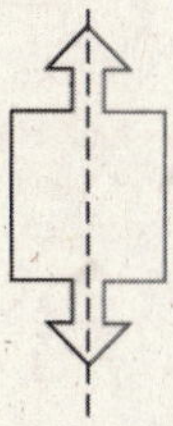

2. 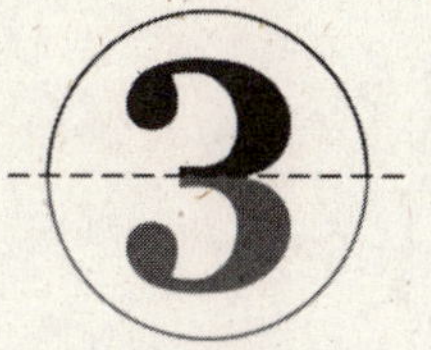

3. 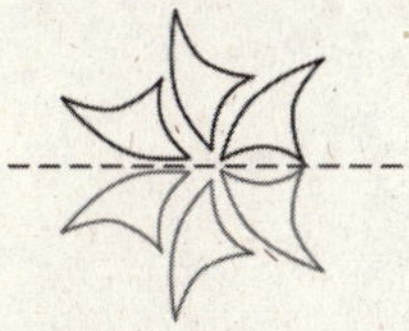

4. 

5. 

6. 

7. Possible answer: 

9. Possible answer: 

Review for Mastery

1. 

2. 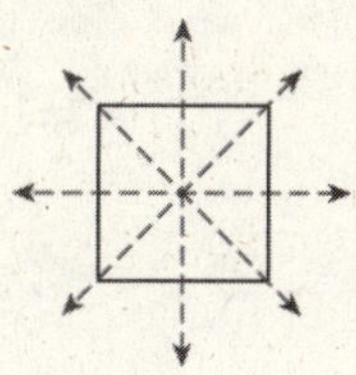

3. 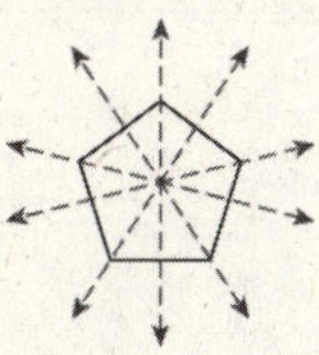

4. 2-fold

5. 3-fold

6. 4-fold

Challenge

1. b. 2
2. b. The line appears on both sides of the strip; one surface

   c. 1 loop

3. 2 loops of unequal lengths linked together
4. 2 loops of same length linked together; one loop is twice as wide as the other
5. 1 loop with a knot

Problem Solving

1. 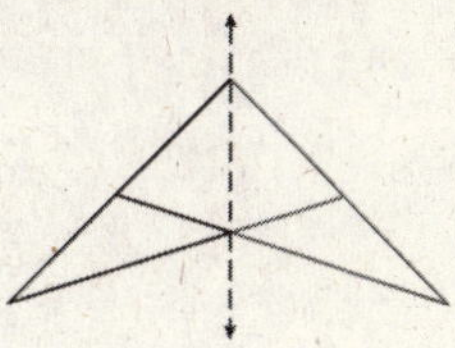

2. 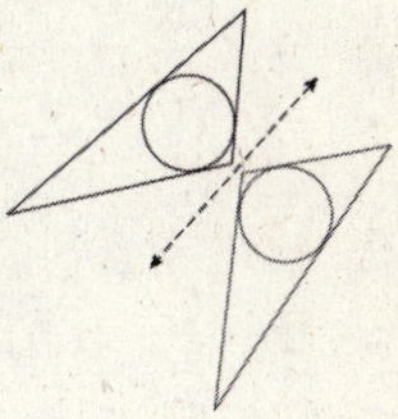

3. 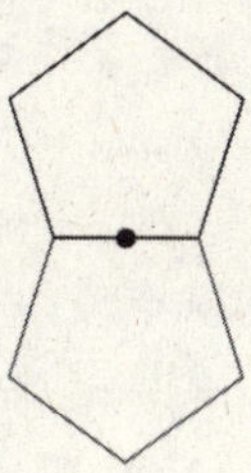

4. 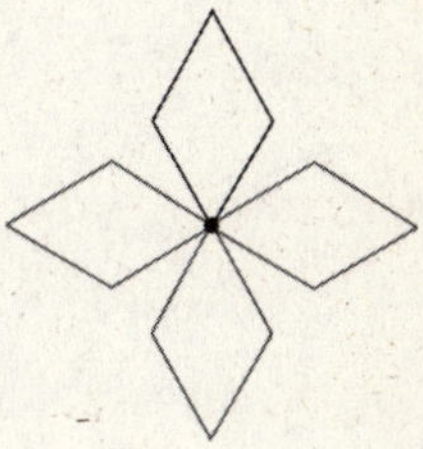

5. D
6. H
7. B
8. H
9. A

Reading Strategies

1. yes
2. no
3. yes
4. no

Puzzles, Twisters & Teasers

Pairs that form a figure with symmetry:
I (first row) and V (last row);
R (first row) and I (last row);
R (second row) and X (third row);
E (second row) and D (fourth row);
T (second row) and A(third row)

Figures without a match:
O (all 3); C (third row); F (third row);
T (fourth row)

A TAXI DRIVER